GENETICS & EVOLUTION

HUMAN GENETICS

GENETICS & EVOLUTION

HUMAN GENETICS

Race, Population, and Disease

RUSS HODGE

FOREWORD BY NADIA ROSENTHAL, PH.D.

☑ Facts On File
An imprint of Infobase Publishing

This book is dedicated to the memory of my grandparents, E.J. and Mabel Evens and Irene Hodge, to my parents, Ed and Jo Hodge, and especially to my wife, Gabi, and my children—Jesper, Sharon, and Lisa—with love.

Facts On File, Inc.
An imprint of Infobase Publishing
132 West 31st Street
New York NY 10001

Library of Congress Cataloging-in-Publication Data
Hodge, Russ.
   Human genetics: race, population, and disease / Russ Hodge; foreword by Nadia Rosenthal.
      p. ; cm.—(Genetics and evolution)
   Includes bibliographical references and index.
   ISBN 978-0-8160-6682-7 (alk. paper)
   1. Human genetics—Popular works. 2. Medical genetics—Popular works. I. Title. II. Series: Genetics and evolution.
   [DNLM: 1. Genetics, Medical. 2. Genetic Diseases, Inborn. 3. Genetic Techniques. 4. Genetics, Population. QZ50 H688h 2010]
   QH431.H515 2010
   599.93'5—dc22 2009010706

Facts On File books are available at special discounts when purchased in bulk quantities for businesses, associations, institutions, or sales promotions. Please call our Special Sales Department in New York at (212) 967-8800 or (800) 322-8755.

You can find Facts On File on the World Wide Web at http://www.factsonfile.com

Text design by Kerry Casey
Illustrations by Dale Williams
Photo research by Elizabeth H. Oakes
Composition by Hermitage Publishing Services
Cover printed by Bang Printing, Brainerd, MN
Book printed and bound by Bang Printing, Brainerd, MN
Date printed: January 2010
Printed in the United States of America

10 9 8 7 6 5 4 3 2 1

This book is printed on acid-free paper.

"I say that it touches a man that his blood is sea water and his tears are salt, that the seed of his loins is scarcely different from the same cells in a seaweed, and that of stuff like his bones coral is made. I say that the physical and biologic law lies down with him, and wakes when a child stirs in the womb, and that the sap in a tree, uprushing in the spring, and the smell of the loam, where the bacteria bestir themselves in darkness, and the path of the sun in the heaven, these are facts of first importance to his mental conclusions, and that a man who goes in no consciousness of them is a drifter and a dreamer, without a home or any contact with reality."

Contents

Foreword

The study of human genetics differs fundamentally from the ways biologists investigate other life forms. Although human experimentation has been largely banned from our culture, the practice of medicine has provided a more detailed description of our species' anatomy and physiology than of any other organism on earth. Vast amounts of observational data embedded in patient medical records lend insights into normal human variation and document the causes and courses of our diseases. Medical research in human genetics is now beginning to yield clinical advances that constitute a revolution. It promises to answer questions about our essential human nature, explain our diseases, and lead to effective treatment. Understanding the genetics of human life is therefore directly relevant to each of us as individuals.

Human Genetics by Russ Hodge vividly records with engaging explanations and captivating anecdotes how the study of human genetics has its roots deep in our heritage as a species but has recently undergone an explosion with the sequencing of the human genome. The goal of his narrative is to introduce students to the history of human genetics, and recent progress in the field, illustrating how its applications to medicine, forensic analysis, and genetic counseling have changed our society. In chapter 1, he describes how the advent of molecular genetics has allowed us to follow the behavior of genes and their interaction with the environment, not only in cells or in tissues but also within families, communities, and cultures. Mutations in genes often have a serious impact on our health, and by studying how they perturb the body and its development, scientists are able to deduce how our systems normally function. Chapter 2 digs deep into our collective genetic past, chronicling how we evolved from remote ancestral organisms to the present day, leaving tantalizing clues to mysteries in our heritage that are only now being solved using molecular genetic techniques, as outlined in chapter 3.

Advances in human genetics also allow clinicians to decipher the genetic basis of rare and common human traits and to integrate this knowledge into the diagnosis, treatment, and prevention of diseases. Chapter 4 describes the goals of current research including gene therapy, which aims to cure disease in patients without affecting inherited traits. Despite concerns about the potential risks of gene therapy—such as changes to the reproductive system and the types of undesirable side effects seen in drugs—this area of medical research holds great promise for many people suffering from the negative consequences of gene defects, and paves the way for individualized medicine.

Genetics, by its very nature, often involves the study of abnormality. This implies that it is possible to define what is normal; since that is so difficult, researchers find it more useful to look at individuals in terms of variation, rather than by comparing them to an abstract "norm." One of the most common misconceptions about our genes is that they determine all human characteristics, which can lead to discrimination against people with certain gene combinations or genotypes. As described in chapter 5, a single genetic mutation can occasionally predict a human trait. But in most cases an individual is a unique, complex product of a huge number of interactions between genes, prenatal conditions, the environment, and lifestyle.

It is essential to fully appreciate these complexities and their implications when we consider how genetic engineering might be used to improve our lot as humans. One day in the very near future, it will be possible to intervene in our own evolution. Doing so would involve a choice: selecting some genetic traits as more desirable than others. The issues surrounding this topic are both moral and technical: Much more will have to be learned about the relationships between genes and the environment before we can effectively alter our hereditary material, should people decide that it is desirable to do so. The book concludes by raising these fascinating and important ethical questions, providing readers the opportunity to explore the issues thoughtfully—a necessary basis for taking a personal stand on the applications of genetic engineering. Human genetics is already personal: It intersects our lives nearly every time we take a medication, undergo a test at a hospital, or check the health of a baby in a mother's womb. The direct impact of this field on each of our lives will continue to grow. We must all be aware of what it means before today's possibilities become tomorrow's reality.

—Nadia Rosenthal, Ph.D.
Head, European Molecular Biology Laboratory
Rome, Italy

Preface

In laboratories, clinics, and companies around the world, an amazing revolution is taking place in our understanding of life. It will dramatically change the way medicine is practiced and have other effects on nearly everyone alive today. This revolution makes the news nearly every day, but the headlines often seem mysterious and scary. Discoveries are being made at such a dizzying pace that even scientists, let alone the public, can barely keep up.

The six-volume Genetics and Evolution set aims to explain what is happening in biological research and put things into perspective for high school students and the general public. The themes are the main fields of current research described by four volumes: *Evolution, The Molecules of Life, Genetic Engineering,* and *Developmental Biology.* A fifth volume is devoted to and titled *Human Genetics,* and the sixth, *The Future of Genetics,* takes a look at how these sciences are likely to shape science and society in the future. The books aim to fill an important need by connecting the history of scientific ideas and methods to their impact on today's research. *Evolution,* for example, begins by explaining why a new theory of life was necessary in the 19th century. It goes on to show how the theory is helping create new animal models of human diseases and is shedding light on the genomes of humans, other animals, and plants.

Most of what is happening in the life sciences today can be traced back to a series of discoveries made in the mid-19th century. Evolution, cell biology, heredity, chemistry, embryology, and modern medicine were born during that era. At first these fields approached life from different points of view, using different methods. But they have steadily grown closer, and today they are all coming together in a view of life that stretches from single molecules to whole organisms, complex interactions between species, and the environment.

The meeting point of these traditions is the cell. Over the last 50 years biochemists have learned how DNA, RNA, and proteins carry out a complex dialogue with the environment to manage the cell's daily business and to build complex organisms. Medicine is also focusing on cells: Bacteria and viruses cause damage by invading them and disrupting what is going on inside. Other diseases—such as cancer or Alzheimer's disease—arise from inherent defects in cells that we may soon learn to repair.

This is a change in orientation. Modern medicine arose when scientists learned to fight some of the worst infectious diseases with vaccines and drugs. This strategy has not worked with AIDS, malaria, and a range of other diseases because of their complexity and the way they infiltrate processes in cells. Curing such infectious diseases, cancer, and the health problems that arise from defective genes will require a new type of medicine based on a thorough understanding of how cells work and the development of new methods to manipulate what happens inside them.

Today's research is painting a picture of life that is much richer and more complex than anyone imagined just a few decades ago. Modern science has given us new insights into human nature that bring along a great many questions and many new responsibilities. Discoveries are being made at an amazing pace, but they usually concern tiny details of biochemistry or the functions of networks of molecules within cells that are hard to explain in headlines or short newspaper articles. So the communication gap between the worlds of research, schools, and the public is widening at the worst possible time. In the near future young people will be called on to make decisions—large political ones and very personal ones—about how science is practiced and how its findings are applied. Should there be limits on research into stem cells or other types of human cells? What kinds of diagnostic tests should be performed on embryos or children? How should information about a person's genes be used? How can privacy be protected in an age when everyone carries a readout of his or her personal genome on a memory stick? These questions will be difficult to answer, and decisions should not be made without a good understanding of the issues.

I was largely unaware of this amazing scientific revolution until 12 years ago, when I was hired to create a public information office at one of the world's most renowned research laboratories. Since that time I have had the great privilege of working alongside some of today's greatest researchers, talking to them on a daily basis, writing about their work, and picking their brains about the world that today's science is creating. These books aim to share those experiences with the young people who will shape tomorrow's science and live in the world that it makes possible.

Acknowledgments

This book would not have been possible without the help of many people. First I want to thank the dozens of scientists with whom I have worked over the past 12 years, who have spent a great amount of time introducing me to the world of molecular biology. In particular I thank Volker Wiersdorff, Patricia Kahn, Eric Karsenti, Thomas Graf, Nadia Rosenthal, and Walter Birchmeier. My agent Jodie Rhodes was instrumental in planning and launching the project. Frank Darmstadt, executive editor, kept things on track and made great contributions to the quality of the text. Sincere thanks go as well to the production and art departments for their invaluable contributions. I am very grateful to Beth Oakes for locating the photographs for the entire set. Finally, I thank my family for all their support. That begins with my parents, Ed and Jo Hodge, who somehow figured out how to raise a young writer, and extends to my wife and children, who are still learning how to live with one.

Introduction

The University of Padua in northeastern Italy is home to some great historical treasures, including a massive wooden lectern from which Galileo Galilei gave courses in mathematics and astronomy and a curious amphitheater where physicians performed autopsies on human corpses. Medical students crammed into the small, lamp-lit gallery and watched from above as a professor dissected cadavers on a table. The body was strapped down so that if the authorities came by for an inspection, the table could be quickly flipped over. On the underside was a partially dissected animal, also strapped down, and the lesson would instantly switch to a discussion of the anatomy of pigs. It was technically legal to study the human body in Padua, although religious leaders protested and it made the authorities uneasy; no one would have been surprised by raids or arrests.

What does it mean to be human? Until very recently in history, this question "belonged" to theologians and philosophers. The body was considered a sacred vessel; nearly everywhere, dissections and most other types of research on human subjects were outlawed. But the Renaissance saw the birth of a new scientific spirit that quickly swept across the Western world. It reached a high point in the mid-19th century, when a series of revolutions made *Homo sapiens* very much the object of research. Human beings were suddenly regarded as collections of cells that grew from a single egg; as the products of evolution and the relatives of all other living things; as a sum of traits inherited from their parents; as walking bags of chemicals, some of which could be synthesized artificially in the laboratory; and as hosts for deadly microorganisms. All of these views of life were invented within just a few decades between about 1830 and 1880.

In the intervening 150 years each of these sciences has developed a very deep and sophisticated perspective on human beings.

Of course an amazing amount remains to be learned in each field, and the different views of human nature that they provide are not yet truly unified. Nor will the picture be satisfying until researchers have a much better understanding of the brain, including phenomena like memory and learning, consciousness, dreams, and spiritual experiences. Even so, there is something remarkable about science at the dawn of the 21st century: Genetics, evolution, cell biology, embryology, chemistry, and medicine are beginning to provide a unified view of the biological side of human existence. And that is the subject of this book.

Every human being carries, within each of his or her cells, a long history of the species. The human *genome* is a record of evolution that stretches back to the first *Homo sapiens* and beyond, to the earliest primates, the first animals, and the origins of life itself. Human genetics is the study of that information and its relationship to people's lives—how their bodies develop, how they behave, whether they are healthy or sick, and other aspects of human existence in which genes play a role. It is the science of how *genes* are passed from one generation to the next. It is also the study of where human *DNA* came from and how it is changing over time.

Most of what researchers know about human genes has been learned from laboratory organisms such as flies and mice, as well as single cells such as yeast or bacteria. Understanding human nature requires comparing different people but also comparing them to other species. One of the great themes of human genetics is to explain why two humans are so similar but unique, and why humans and chimpanzees are so alike and yet so different. Like most topics in modern biology, these studies only make sense because of evolution: Organisms have inherited nearly all of their genes and bodybuilding processes from their common ancestors. So evolution and studies of other species are recurrent themes of this book.

Most books about human genetics are written for college students or experts. But the themes of this field need to be widely known. Even seemingly trivial discoveries in this area of science have a way of suddenly turning into applications that affect many people and society as a whole. DNA fingerprinting, which is used every day to solve crimes and resolve questions of paternity, arose from a study of the evolution of a protein found in muscle cells. Genetic engineering experiments with mice reveal connections between genes and disease that may lead to new prenatal tests and difficult decisions for families expecting babies. So everyone needs to understand the basics of a science that will have an impact on his or her personal health and society as a whole over the next few decades.

What does it mean to be human? *Human Genetics* aims to weave together several perspectives. Chapter 1 looks at human beings as individuals that arise through an interplay of genes and the environment. Chapter 2 looks at the entire species as a product of the changes that have occurred in the genome. Studies of human molecules have been applied in some fascinating ways—for example, to solve historical mysteries—and these are the subject of chapter 3. These themes come together in medicine as modern doctors try to identify the factors that make the body healthy or sick, which is the topic of chapter 4. The medical applications of genetics promise to change all our lives. Finally, chapter 5 looks at the rich variety of the human species—differences between individuals and groups, including questions like the genetic meaning of human races, and how genes influence behavior and society.

Human genetics, like the rest of today's biology, is a materialist science; it regards the body and living processes as the result of chemical and physical laws. But one thing it has revealed is that genes do not function like computer programs, set running and left to crunch numbers while a programmer catches some sleep. What happens in cells and organisms is always the result of a dialogue with the environment. The human environment includes the weather and the natural landscape, the smog above a city, the cubicle where a person works, cramped airplane seats and the chemicals in drinking water—but it also includes human society, science, the Internet, music, and ideas. British biologist Stephen Rose put it this way: "I don't think we can understand what it means to be human without understanding that we are evolved organisms, just as much as we are social, historical, cultural and technological organisms. To understand human nature means that we have to understand all of these things."

A child inherits a body from his or her parents, along with a few unique features, but this body does not come with a user's manual. Creating such a manual for a machine requires a thorough technical description of its parts, their functions, and how they work together. Today's biology is providing such a description of the body. But the manual itself can never be written by science; that is the job of every individual. This book aims to be a good place to start.

1

From Genes to Human Beings

A doughnut-shaped red blood cell and a neuron in the brain look so different from each other that it is hard to believe they come from the same organism or even the same species. But the two types of cells are the product of a single "recipe book." They arise from the same genome but use it in different ways as they grow and develop. The same set of information produces hundreds of types of cells with distinct appearances and behavior. That information is created when an egg is fertilized; it reproduces and differentiates until there are about 100 trillion cells, which work together to create a human being.

Genetics began as a study of patterns of heredity, usually in adult plants and animals. That is still an aim of population genetics, which studies how the genes of a group change from generation to generation, and hereditary patterns are essential in the search for genes related to disease. But as scientists have learned what genes are made of and how they function, the focus has shifted a bit. The genome encodes a dynamic set of processes that begin in the chemistry of DNA sequences and produce the features of organisms. This chapter briefly introduces how modern genetics arose and presents the basic concepts needed to understand its relevance to human beings.

MENDEL AND THE LAWS OF HEREDITY

Until the second half of the 19th century, people had only a superficial knowledge of heredity. The ancient Jews had observed that

1

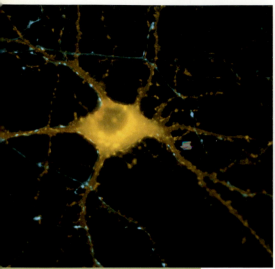 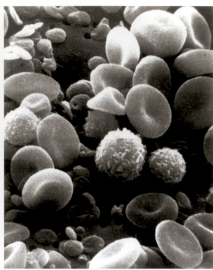

The same genome produces a wide range of cell types, including treelike neurons and blood cells in many shapes and sizes. *(Michael A. Colicos, Division of Physical Sciences, USCD)*

hemophilia, a deadly disease in which the blood fails to clot, ran in families and could be transmitted from mother to son. While each newborn boy was required to be circumcised, an exception could be made if his mother's family had a history of the disease. Farmers and breeders knew that crops and animals could be changed by collecting the seeds of plants with desirable qualities, or by allowing specific animals to mate. But until late in the 19th century, knowledge of the process by which organisms passed traits on to their offspring was patchy and confused, as illustrated by the story of a young Englishwoman named Mary Toft. In 1726 she tricked some of the most prominent doctors of her day into believing that she had given birth to rabbits. The fraud was eventually exposed, but the fact that many people were taken in reveals how little was really known about heredity.

The 18th century saw a few attempts to study human heredity in a scientific way. The French mathematician and philosopher Pierre-Louis Maupertuis (1698–1759) tracked the appearance of extra fingers through four generations of a family—the first known description of a genetic disorder in humans. He also constructed family trees of people with *albinism* (a condition in which people lack pigment in their skin, hair, and eyes) and investigated the inheritance of color patterns in dogs. A half-century later, Joseph Adams (1756–1818) wrote *A Treatise on the Supposed Hereditary Properties of Diseases,* in which he recognized that marriages between close relatives often produced unhealthy offspring. But these remained anecdotes until the latter half of the 19th century, when Johann Gregor Mendel (1822–84) began to investigate the problem of heredity very systematically. His experiments allowed him to detect a set of laws that underlie patterns of inheritance.

Mendel spent most of his life in an abbey in Moravia, now part of the Czech Republic. Today this might seem like an unlikely birthplace for a new science, but 19th-century abbeys were often centers of learning and research, and many clergymen were amateur scientists. Often entering a monastery or cloister was the only way for a young man or woman of the lower or middle classes to gain an education. The alternative for Gregor Mendel was to work on his family's farm, but health problems made him unfit for the strenuous life of a farmer. He was a promising student and his sister paid his way to the university. But during his studies he suffered from severe nervous attacks whenever he had to take tests. One of his professors suggested that he join an abbey, and recommended Saint Thomas in the town of Brno. The abbey offered Mendel a sheltered life away from stress of the "real world." His nervous condition prevented him from taking on a congregation as a priest. But Saint Thomas's abbot had taken a liking to the young man, who adopted the name Gregor, and allowed him the leisure to pursue scientific activities while living in the abbey.

Mendel devoted himself to one of the most intriguing scientific problems of his time. He loved mathematics and statistics and found a way to use them as tools to study the patterns by which traits were passed from parents to their offspring. He began working with mice, but he had to give up that project when a visiting bishop objected to finding animals mating in a monk's room. The abbot gave him a sizeable plot in the monastery garden where he could work with plants. What at first seemed like a setback eventually made an important contribution to Mendel's success. Animal heredity is in many ways more complex, and it is unlikely that Mendel would have achieved the same results if he had continued working with mice.

Other scientists of Mendel's day, including Charles Darwin (1809–82), were conducting experiments to uncover the principles behind heredity, but only Mendel succeeded, for several reasons. First, he took an original approach to the question of heredity. A plant or animal is the sum of many features, and his first decision was to examine the inheritance of each feature separately. This was an important step. No one knew whether features were inherited together—as if parents gave their children a finished "stew" of traits—or separately, which would be like passing on the ingredients and recipe needed to make a stew. Mendel began by assuming that traits were inherited separately, and the strategy allowed him to show that this was indeed the case.

Another important factor was the care with which he designed his experiments. Mendel's original study focused on peas. He decided to investigate several characteristics that were easy to observe and hard to misinterpret, including the shapes of peas (round or wrinkled), their colors (green or yellow), the colors of pods, and the length of the plant's stems. He spent two years getting ready, breeding the plants over and over again until they consistently produced the same features. This gave him a "parent generation" of plants with predictable characteristics. He could now mate different types with each other to study which features of each parent were inherited by the next generation.

While carrying out this work it was necessary to prevent contamination, which could arise from a transfer of pollen between plants through accidents or insects. The male and female reproductive organs of peas are near each other; in the wild, a plant usually pollinates itself unless an insect transfers pollen from another plant. Mendel controlled this process by cutting off the male organs of the plants, called anthers, and wrapping the female organs in small bags. To breed the plants he selected a particular type of father plant, removed pollen from it, and transferred it to the mother. This painstaking methodology was equally important to success; contamination would have made all of his results useless.

One unique aspect of the study was the fact that Mendel tracked features over more than one generation. Traits such as blue eyes often skip a generation; the children of a blue-eyed father and brown-eyed mother may all have brown eyes, but blue eyes may reappear in the grandchildren's generation. This phenomenon is crucial in understanding heredity, but it can only be observed by following many generations. If Mendel had broken off his experiments after producing a first generation of peas—as many other researchers had done—the patterns would have been incomprehensible.

Mendel began by taking pollen from his wrinkled-pea strain and using it to fertilize plants with round peas. In a second experiment he introduced round-pea pollen into wrinkled-pea strains. The plants produced several hundred seeds (called "first-filial-generation" hybrid seeds, or F1) that were all round. A year later Mendel planted the F1 seeds, this time allowing the plants to fertilize themselves. The F1 plants produced 7,324 second-filial-generation peas (F2), of which 5,474 were round and 1,850 were wrinkled. This gave a proportion of 2.96:1 in the seeds of the second generation—nearly three to one. It made no difference which parent had contributed which element—the proportions turned out virtually the same. This meant that the two sexes contributed equally to the characteristics of the offspring.

The three-to-one ratio led Mendel to several brilliant insights. He realized that each of the features he was studying (for example, wrinkled versus roundness) was composed of two "elements"—one inherited from each parent. One element was *dominant* and the other type was *recessive,* meaning that if a pea had one element of each type, it would take on the dominant form. This explained why the first-generation peas were all round: Each had inherited a round element from one parent and a wrinkled from the second.

The principle of dominant and recessive traits also explained what happened to the second generation, when the F1 plants fertilized themselves. Each of their offspring received a chance combination of two traits. Statistics allowed Mendel to predict that one fourth of the plants had received two round elements (producing round peas); another fourth had inherited two recessive, wrinkled elements (making them wrinkled); and two-fourths had received one element of each type (also turning out round, because this shape was dominant). Mendel confirmed the pattern by studying other traits in peas and other plants.

A series of unfortunate events meant that Mendel never received widespread recognition for his work during his lifetime. The results were published in a small scientific journal but went virtually unnoticed. Mendel had been corresponding with a prominent botanist, Carl von Nägeli of the University of Munich, who was initially helpful, but ultimately the relationship did more harm than good. Nägeli had been experimenting on another plant, the hawkweed, which was difficult to work with. Mendel's rules did not explain the patterns of inheritance that Nägeli had found. Mendel began an intensive study of the hawkweed, but he was equally unable to make sense of the results. Unknown to either man, the plant has an unusual form of reproduction in which plants sometimes arise from unfertilized eggs. Of course this skewed the number of traits seen in every generation. Mendel became frustrated and published a partial retraction of his earlier results. When he was asked to become abbott of the Saint Thomas monastery, he accepted and began devoting most of his time to administrative duties. Fifteen years after his death, his work was rediscovered and became the basis of a new science called genetics.

CELLS, CHROMOSOMES, AND SEX

Mendel's work produced the concept of a unit of inheritance that contained the information for a single feature of the organism, passed from parents to their offspring. (Mendel called them "units"; the term *gene* was proposed decades later by the Danish researcher Wilhelm Johannsen.) Had anyone asked Mendel to find one of his units in a cell, he would have been unable to do so. This remained the case for nearly half a century, despite the intensive research into genetics that followed the rediscovery of Mendel's work. A first step toward resolving the question came at the beginning of the 20th century, when scientists proved that genes were located on chromosomes. This was confirmed over the next few decades and was an important step in understanding the chemical and physical nature of heredity. The discovery of the double-helix structure of DNA proved once and for all that genes were made of DNA.

Traits obviously had to be transmitted from parents to their offspring through a material substance—something in the fluids exchanged during sex. But finding this substance would require a close look at the structure and chemistry of cells. Fortunately, as the science of genetics was born, cell biology was undergoing a revolution of its own. New types of microscopes and dyes were giving scientists their sharpest view ever of the inner world of cells. In 1840 this permitted the young Germans Matthias Schleiden (1804–81) and Theodor Schwann (1810–82) to claim that all plants and animals were made of cells. Their work led to a new theory, proposed 15 years later by their countryman Rudolf Virchow (1821–1902), that every cell arises from a pre-existing cell. An embryo began as a single fertilized egg that divided over and over again. Each time it had to pass hereditary information on to its daughters. But where was that information located, and what was it made of?

At the end of the 19th century scientists began to focus on the cell nucleus. In 1876 Oskar Hertwig (1849–1922) was the first to watch the fusion of a sperm and egg under the microscope, using the huge, pearly-white egg cells of sea urchins. He discovered that the entry of a sperm brings a new nucleus into the egg, which then fuses with the egg's own nucleus. The rest of the sperm is broken down and disappears, meaning that the hereditary material of the father had to be contained in the nucleus. Its fusion with the nuclei of the egg produced a new organism with unique characteristics.

New types of dyes gave microscopists their first look at the contents of the nucleus. Walther Flemming (1843–1905) discovered *chromosomes,* threadlike structures that appeared and disappeared during different stages of the cell's life cycle. (The DNA strand is far too thin to be seen under the microscope, so chromosomes only appear when DNA is packed into huge, tight bundles—such as when the cell is about to divide.) In 1879 Flemming noticed that chromosomes were split up into two sets when the cell divided. Wilhelm Roux (1850–1924) and August Weismann (1834–1914) observed how they were combined again during fertilization. Chromosomes, Roux wrote, must contain the hereditary material, and he proposed that the information they contained was in a linear form, like the words of a text. In 1900 Theodor Boveri (1862–1915) proved that different chromosomes are responsible for different hereditary characteristics.

That same year, three scientists working independently rediscovered Gregor Mendel's work while carrying out their own research into heredity in plants. Hugo de Vries (1848–1935), Carl Correns (1864–1933), and Erich Tschermak von Seysenegg (1871–1962) had approached the question of heredity much the same way Mendel had and arrived at the same basic conclusions. De Vries alone studied and confirmed Mendel's laws by observing patterns in about 20 species of plants.

At the same time British researcher William Bateson (1861–1926) had been working on heredity in plants and animals. His experimental method did not lead him to reproduce Mendel's results, but de Vries's articles pointed him to the monk's experiments. Bateson immediately realized that some of the major questions of heredity had been solved. He translated Mendel's original article into English, wrote a book explaining its meaning for science, and made sure that Mendel's name became known throughout the scientific world. In the process he realized that the new science needed a new vocabulary and invented some of the key terms of modern genetics. He coined the term *allele* to refer to variants of an existing gene. For example, peas had a gene that controlled their shape. There were at least two alleles—one for roundness and one for wrinkles. If an organism inherited a copy of each, the allele for roundness was dominant.

The discovery of sex chromosomes, largely through the efforts of Walter Sutton (1877–1916) and Nettie Stevens (1861–

1912), was another important step in establishing the role of chromosomes in heredity. Sutton had been trained at the University of Kansas in the laboratory of Clarence McClung (1870–1946), who shared Boveri's hypothesis that each chromosome contained a subset of a plant's or animal's hereditary material. Working with grasshoppers and other insects, Sutton showed that when chromosomes "reappeared" at the beginning of cell division, they formed the same shapes they had had before disappearing. That might make it possible to determine which chromosome was responsible for an animal's sex. Originally Sutton believed that an additional, "accessory" chromosome, which he called the X chromosome, was responsible for making the egg into a male.

Nettie Stevens, one of the few women to receive an advanced degree in science in turn-of-the-century America, soon convinced him that he had it backward. Stevens had received her Ph.D. at Bryn Mawr with geneticist Thomas Hunt Morgan (1866–1945), followed by a year of work abroad with Theodor Boveri. It was excellent preparation for tackling the problem of the inheritance of sex. Working with mealworms, she discovered that females had 20 large chromosomes, whereas males had 19 large chromosomes and one smaller one. Sperm with 10 pairs of large chromosomes produced females; those with nine and the small 10th chromosome ("Y") became males. At Columbia University, Edmund Beecher Wilson (1856–1939) was finding the same phenomenon in the chromosomes of several species of insects. They had solved one of history's greatest mysteries: the cause of the difference between the two sexes in human beings and a wide range of other species.

Yet there were exceptions to the rule. There are rare cases of women who have only one X chromosome, and a few males have two Xs as well as a Y. This meant that something about the Y made an embryo into a male—rather than the fact that one X was missing—but what was it? The answer remained a mystery until the early 1980s, when Robin Lovell-Badge of the National Institute for Medical Research, working with Anne McLaren and Paul Burgoyne of the Medical Research Council (all in London), discovered a gene called SRY. Normally found on the Y chromosome, SRY had to be present for mice to become male. Soon they showed that the gene plays a similar role in humans. In very rare cases, SRY jumps onto an X chromosome and is found in embryos with two Xs, and this turns what would otherwise be a female into a male. At other times the egg inherits the XY pair but the SRY gene is missing, which leads to a female embryo.

The discovery had a completely unexpected consequence. For several years the International Olympic Committee had carried out medical inspections to ensure that the participants in women's events were really females. Now it seemed the SRY gene could be turned into a genetic test for gender, and it was put to use for the 1996 Summer Olympics event in Atlanta, Georgia. Several females tested positive for the gene, but the test was flawed, so no one was barred from competing. After a protest by several American medical associations, genetic testing for gender was dropped again in 2000. Since then, other genes on the Y chromosome have been found to contribute to the development of male sex characteristics.

DISCOVERING AND MAPPING GENES

How many genes does it take to make a plant, animal, or human being? Today scientists estimate that a human cell probably encodes between 20,000 and 25,000 genes, about three times the number found in a yeast cell and less than twice the number found in a fly. These figures only became available in the first few years of the 21st century, after scientists had learned a great deal about the chemistry and functions of genes and obtained complete genome sequences from the organisms. Yet a century ago, even without knowing that genes were made of DNA, the laboratory of Thomas Hunt Morgan made amazing progress in identifying new genes and describing their characteristics and functions. The findings of Morgan and his colleagues have had an extremely important impact on the development of human genetics, laying the groundwork for the discovery of new genes and mutations linked to disease.

Morgan's career as an independent researcher began when he was hired by E. B. Wilson, who was setting up one of the first biology departments in the United States at Bryn Mawr. Wilson moved on to Columbia to set up a similar department there, and Morgan followed him a few years later. As he set up his new laboratory he began looking for an animal that was easy to care for and breed in large numbers. A colleague recommended *Drosophila melanogaster,* the fruit fly, known to swarm around moldy fruit. It reproduced within two weeks of birth, could be kept in glass jars, and needed only a diet of mashed bananas.

Morgan hoped that his studies of flies would allow him to "catch evolution in action." Scientists were unsure how to link the new science of genetics to the theory of evolution. Part of the problem was that the two fields asked different types of questions, and no one was sure whether—or how—the questions overlapped. Mendel and the early geneticists were mainly concerned with the way existing genes were shuffled around in a population—if peas could be round or wrinkled, which trait would be passed from a parent to its offspring? Evolution's main concern was how the existing features of a population changed to produce new species.

Somehow the two processes had to be connected. Natural selection could explain why natural forces might help one allele survive and another be eliminated (for example, round peas might float and survive a flood, whereas wrinkled ones might sink). But evolution needed something more. New species did not simply arise simply by mixing up a species' current set of genes. Existing genes had to undergo changes and new ones had to arise. Hugo de Vries had developed a hypothesis that genes could undergo changes that he called *mutations.* If these changes were passed along to an organism's offspring, they could become the stuff of evolution. Morgan hoped to observe such changes in fruit flies and follow their effects on evolution.

In January 1910, after two years of unsuccessful searching, Morgan's lab witnessed the first mutations in flies—a small change in the coloring of their bodies. Soon he found a much more dramatic example: a fly whose eyes were white instead of the normal red. When the fly was mated with others, its white eyes were passed down to some of the offspring. Morgan interpreted this to mean that flies had a gene for eye color that had undergone a mutation. He began a tradition of naming genes after the effects of mutations. So he called the newly discovered gene "white," even though the normal form of the gene produces red eyes.

More discoveries followed quickly. Within a short time, Morgan's lab discovered genes responsible for dozens of features of the fly's body—from its size to the shape of its wings. His focus quickly began to shift from evolution to questions about the puzzling behavior of genes. Some of the patterns did not seem to fit Mendel's rules. Crossing red-eyed males with white-eyed females, for example, led to sons with white eyes and daughters with red. But switching the roles—crossing white-eyed fathers and red-eyed mothers—produced a first generation that all had red eyes. Their male offspring showed a 3:1 ratio of red to white. Mendel's rules were working, but the sex of the fly was somehow skewing the pattern.

Morgan's colleague E. B. Wilson solved the mystery by proposing that the gene for eye color might be located on the X chromosome. Females had two copies of any gene located there (because they had two X chromosomes) and males only one, which they inherited from their mothers. Somehow this gave males a trait that did not appear in females. Wilson suddenly realized that the same phenomenon might explain something he had observed in human inheritance. He was extremely color-blind and had been studying how this was inherited in families. His genealogies suggested that this trait, too, was passed from mothers to sons. The same turned out to be true of hemophilia and many other genetic diseases. The responsible genes might be located on the X chromosome. Compiling a list of such traits ought to tell researchers which genes were located there.

When the sex of the fly was taken into account, Morgan discovered that the inheritance of traits such as white eyes followed Mendelian patterns—in other words, one gene was responsible for the change. This added to Mendel's theory by demonstrating that genes could be damaged through mutations; the same process might produce new alleles.

Next, the researchers discovered that traits were not inherited completely independently. A newborn fly might inherit white eyes because its mother had donated an X chromosome with the mutant form of the white gene. At the same time, it inherited the other genes on the X chromosome. The laboratory kept precise statistics on the new strains of flies and the results of experiments in which they were crossed with each other. X chromosome traits were usually inherited together, while genes located on different chromosomes were usually inherited independently. This was not hard to understand. During the creation of new egg or sperm cells, a fly split up the pairs it had inherited from its own parents. They were split randomly, so an egg might receive a copy of chromosome 1 that had come from the fly's father, chromosome 2 from its mother, and so on in a random way.

But genes on the same chromosome were not always inherited together; in rare cases two genes that had always been linked suddenly began behaving independently. At other times two independent genes became linked. These discoveries were only possible because of the huge number of flies in the lab and careful accounting.

In 1912 Morgan and Alfred Sturtevant (1891–1970), a member of the lab, came up with a possible explanation. Before the chromosomes are separated in the production of eggs and sperm, they are lined up and twisted tightly around each other. Morgan thought that when bent so sharply, there might be breaks in chromosomes. Cells could repair the breaks, but when this happened, genes might change positions.

If this was the case, Sturtevant said, it offered a way to make maps of the positions of genes on chromosomes. Breaks in chromosomes were more likely to split apart genes that were located far from each other than those that were close together. The situation is a bit like a puzzle that has been taken apart in a hurry, leaving blocks of pieces connected to each other. Adjacent or nearby pieces are more likely to be found in the same block than those that are far apart in the completed puzzle. This principle, called *linkage,* allowed Sturtevant to chart the relative positions of several genes on a chromosome. By analyzing the probability that two traits were inherited together when different strains were mated, he could tell which genes were closer to each other on the chromosome. Eventually the method allowed the laboratory to map dozens of genes to specific locations.

The same basic method is used today to search for connections between genes and human diseases, although there are simpler ways than looking for linkage between separate genes. Instead, DNA sequences called *microsatellites* are used to compare patterns of inheritance in healthy people versus those affected by a genetic disease. This topic is covered in detail in chapter 3.

FROM GENOTYPE TO PHENOTYPE

When an egg and sperm cell fuse, DNA from both parents is brought together to create the genome of a new organism. Genes carry the information needed to build a complete human being. Sometimes the information is defective and causes problems as the embryo develops. A change in a single letter in the genetic code may make the difference between a healthy person and an embryo that dies or suffers severe health problems. To learn why this happens, scientists needed to understand how an organism's *genotype* (its complete set of hereditary information) produces its *phenotype* (all the characteristics of its body and behavior influenced by its genes). The first step in understanding this process was to learn how the information in a single gene was used. Answering that question was the top priority of molecular biology laboratories from the 1950s to the 1970s.

Thomas Morgan's laboratory was content to regard genes as abstract ideas, like variables in an algebra equation. In the early 20th century scientists did not know what kind of molecule genes were made of—many thought that *proteins* contained the genetic code. But in 1953 the young American James Watson (1928–   ) and the Englishman Francis Crick (1916–2004) discovered the structure of DNA. The chemistry and double-helix form of the molecule proved that it could contain the hereditary material. The structure also revealed how cells could copy their DNA to pass it along to their offspring and how mutations might arise.

It did not show, however, how the information in the molecule was used to make different types of cells and build organisms. Genes were not only a sort of reference library; they also played an active role in the daily life of the cell. Experiments in the 1940s by George Beadle (1903–89) and Edward Tatum (1909–75) showed that a mutation in a single gene caused a type of mold to lose a single enzyme (a type of protein). This meant that each gene was responsible for the production of one protein. The principle helped explain how genes had their effects on organisms. Proteins are known as the "worker" molecules of the cell. A few of their functions include receiving signals that tell cells how to behave, interacting with each other to manage the cell chemistry, and forming fibers and other structures that give cells their shape.

Even the double helix, however, did not explain how information in DNA could be transformed into proteins. The two were completely different kinds of molecules. The main goal of biology over the next two decades was to show how one type of information could be translated into the other. This was crucial to understanding why mutations caused problems.

Crick outlined an answer in 1958 when he stated what is called the "central dogma" of molecular biology: "DNA makes RNA makes proteins." Genetic information first had to be transcribed into an intermediate molecule called ribonucleic acid (*RNA*), which was then used to make a protein. Crick's statement was a challenge to the entire scientific community to figure out how cells managed the steps in the process. The idea had several implications. First, "DNA makes RNA makes proteins" meant that information was transmitted in a one-way direction. An RNA was produced from the information in a gene, but it could not send information back and change the gene. Likewise, an RNA could be used to make a protein, but a protein could not influence the content of the RNA.

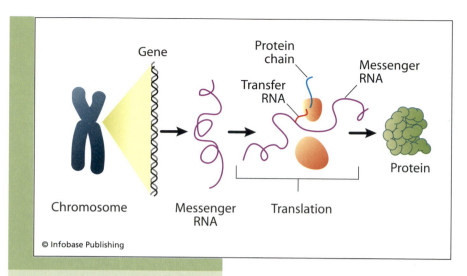

Gene

Protein chain

Transfer RNA

Messenger RNA

Chromosome

Messenger RNA

Translation

Protein

© Infobase Publishing

Information in genes is transcribed into a similar molecule called RNA, which is processed into messenger RNA and then translated into protein by the ribosome.

Since then, researchers have discovered several exceptions to the rule—RNAs can, indeed, sometimes rewrite the information contained in a gene. And Beadle and Tatum's principle of "one gene makes one enzyme" has also proven to be too simple. A gene cannot encode two completely different proteins, but the RNA that is made from it can be cut and pasted together in several ways to produce very different forms of a protein, just as the same recipe can lead to different dishes if a cook leaves out some of the steps. At every step in the transformation of genetic information into proteins, cells have evolved mechanisms that enable them to step in and refine or block the process.

Reading the information in a gene and transforming it into a protein requires the help of dozens—sometimes hundreds—of other molecules. Once it has been made, the protein likely does its job as a part of a "molecular machine" that contains dozens of other molecules. Since different people have slightly different versions of single genes, their machines—and thus their cells and bodies—develop in different ways. Diet, infections, poisoning, and mutations can also have an effect on which genes are used and how proteins function.

This means that the connection between a genotype and phenotype is somewhat flexible and depends on many influences that cannot be foreseen. A mutation may make it more likely for someone to develop cancer or *Alzheimer's disease* late in life, but whether that happens to an individual can rarely be predicted with absolute certainty. What looks like a dangerous form of a gene may be offset by other genes that a person has inherited, his or her lifestyle, or environmental factors. This has important implications in interpreting genetic tests, such as those performed on a fetus to determine whether it might develop a serious problem. It also shows that while genes make human behavior possible, and may strongly influence people's actions, they usually only predetermine a person's fate to a limited degree. Chapters 4 and 5 explore these themes in more depth.

STEM CELLS AND DEVELOPMENT

Building a human body requires the creation of hundreds of different types of cells with very different appearances and behavior. They all arise from a single cell and possess the same genome. In spite of this, they take on unique characteristics because each type of cell uses a different set of its genes. A typical human cell may only use a third or a fourth of its genes. That set determines which proteins it contains, how it is shaped, and what stimuli it can respond to.

A human begins as a fertilized egg that is *totipotent*—it can develop into all of the cell types in the body. When it divides, it produces *embryonic stem cells* that are also totipotent. But very quickly—when the embryo consists of about 32 cells—they begin to specialize. One reason for this is that the huge egg cell copies its nucleus and subdivides for a while without becoming any larger; the newborn cells arise in regions of the original cell that are not identical. An analogy would be the way the former Soviet Union has split into many smaller countries. The unique history, language or dialect, and customs of different regions have influenced the way each new country is developing. Before a fertilized egg divides, various regions of the cell contain different proteins. This creates individual chemical environments that activate different sets of genes in the new cells, causing them to take on unique fates.

Pools of *stem cells* are maintained as the embryo grows; they can be called up to replace damaged tissues. These cells are *pluripotent*: they can still differentiate into many types, such as *hematopoietic* stem cells that specialize into red and white blood cells. But they are no longer able to become every type in the body.

In the third week of a human embryo's life, cell division and specialization lead to the formation of three layers of tissue with unique characteristics. In this process, called *gastrulation,* they slide along each other and fold into each other. The migrations bring cells into contact with new neighbors that produce unique proteins. Encounters between molecules on the surfaces of the cells trigger the activation of new genes. The cells receive instructions that tell them what to become.

The three layers are called the *ectoderm* (outer), *mesoderm* (middle), and *endoderm* (inner layer). These will produce hundreds of specific cell types and the body's organs. Some organs arise from a single layer; others are built from the cells of two or three layers. Even when one layer is the source of an entire organ—most of the brain develops from the ectoderm, for example—this cannot happen without input from the other layers.

Although each kind of cell and organ is unique, a small set of powerful signals is used over and over in the development of many different tissues throughout the body. Defects in many of these signals have also been linked to cancer and other diseases. A signaling molecule called Wnt, for example, helps create cell types and structures in the skin, kidneys, blood, and many other organs. It is also highly active in some types of tumors. This makes sense because of the way that animals evolved. The ancestor of multicellular organisms was a single cell, which means that it was also the ancestor of every type of cell in their bodies. As bodies evolved sophisticated organs, they had to draw on genes and signaling pathways that were already available.

But these processes unfolded differently in various parts of the body because cells produced different molecules to control and interpret the signals.

Organ building requires the creation of new types of cells that reproduce, migrate, and build large structures. In the embryo this often begins when stem cells migrate to a new location, where they encounter a new set of stimuli. In one location this process produces a brain; in another the result is a liver. A tumor arises through a similar process. Animal organs probably originally arose as tumorlike masses of cells that somehow helped the organism and then were refined through millions of years of natural selection. But in the vast majority of cases, tumors disrupt the functions of other organs and are eventually fatal.

Signals that stimulate the formation of an organ need to be switched on and off at the right times. The most obvious reason is that early in development, stem cells divide very quickly to produce an organ of the right size. But tissues—and the body as a whole—cannot simply keep growing. Once it has reached its optimal size, most of an organ's cells have specialized and stop dividing. If for some reason they reactivate an earlier developmental program, they may produce a tumor.

On the other hand, many tissues maintain a small pool of non-specialized stem cells that can be called up to replace cells that have been damaged or worn out. Bone marrow, for example, is home to the hematopoietic stem cells used to make blood. They are needed constantly; many types of blood cells have short life spans and have to be replaced all the time. (Red blood cells, for example, only live for about 120 days.) Molecular signals actively protect stem cells from specializing, because they seem to have a "default" program to specialize. This happens quickly if the cells stop dividing.

Most other adult tissues cannot regenerate themselves very efficiently, probably because there are very few adult stem cells of the right type. In the embryo, however, many tissues can be rebuilt at least partially when they have been damaged, probably because of the presence of high numbers of stem cells and the fact that fetal tissues are still "programmed" for growth.

Much of this ability is lost after birth, and it steadily decreases over the course of a lifetime. By understanding the signals that protect stem cells or cause them to differentiate, researchers may one day learn to retrain a person's cells to replace their damaged neighbors. Such cells could also be used to generate new blood, particularly for people with rare types, and in 2008 researchers achieved a breakthrough in this area (see "Making Blood in the Lab").

The strategy could also be a powerful tool in conditions such as *Parkinson's disease,* in which neurons die in a region of the brain called the substantia nigra. Another possibility might be to transplant stem cells from one person to another, but foreign cells—like foreign organs—are likely to be rejected by the immune system. A final approach is to save and freeze a person's own stem cells. Some hospitals have begun to collect cells during births from the umbilical cord, which usually contains highly potent embryonic stem cells from the newborn, in anticipation of a day when a patient will need them.

Most researchers believe that some of these approaches will be successful, but it may be many more years before they become common tools in the physician's arsenal.

MUTATIONS

Science fiction stories and films such as the X-Men series or the *Heroes* television series often portray mutations as bizarre events that give people superhuman powers or make them into monsters. In reality all human beings are born with mutations that give them unique genomes (with the exception of identical twins, whose genes are identical). While many mutations have no noticeable effect at all, or small effects on the chemistry of the cell, some may influence the entire body by changing the way cells behave and organs develop. They may lead to disease or deformities. On the other hand, they may give someone a slightly better immune system, or improve the way blood transports oxygen, which would be an advantage in sports. But they will not allow a person to break the laws of physics by allowing

Making Blood in the Lab

Hospitals across the world depend on constant efforts to collect blood from donors. Sometimes there are shortages—especially in times of crises, or when it comes to supplies of the rare AB-negative type. One solution might be to cultivate stem cells in the laboratory and stimulate them to develop into blood. If successful, the procedure might have other advantages: The cells could be used in place of bone marrow transplants to treat people suffering from autoimmune diseases such as rheumatoid arthritis and

Robert Lanza, chief scientific officer of Advanced Cell Technology, is a pioneer in stem cell research and cloning. His accomplishments include stimulating embryonic stem cells to develop into blood and cloning endangered species. *(Robert Lanza)*

multiple sclerosis. In a recent breakthrough, researchers from the company Advanced Cell Technology (ACT) in Worcester, Massachusetts, the University of Illinois at Chicago, and the Mayo Clinic in Rochester, Minnesota, have brought the idea much closer to reality.

The project is headed by Robert Lanza, vice president of medical and scientific development at ACT, renowned author, and a globally recognized expert on cloning and regenerative medicine. The work with blood is a

(continues)

(continued)

continuation of years of research on stem cells by Lanza and his colleagues. In an article in the November 24, 2001, issue of *Scientific American,* the team reported the first successful cloning of a human embryo—taking the nucleus from cells called cumulus cells, injecting it into an egg cell, and stimulating it to divide. "In the end it took a total of 71 eggs from seven volunteers before we could generate our first cloned early embryo," the team reported. "Of the eight eggs we injected with cumulus cells, two divided to form early embryos of four cells and one progressed to at least six cells before growth stopped."

The goal of the work was not to allow the cells to develop into an entire human being—even if they had survived, this would have required implanting them in a mother's uterus. Instead, Lanza and his colleagues hoped to use them to obtain embryonic stem cells that could then be raised in laboratory cultures. By stimulating the cells with specific factors, the researchers hoped, they might be able to prompt the cells to develop into specific types of tissues. Making blood was one goal. The strategy might also produce neurons, which could be used to replace cells lost in conditions like Parkinson's and Alzheimer's disease, or pancreatic "islet" cells, which make insulin.

That same year, the methods allowed the team to clone an animal called the gaur, an enormous wild cow that inhabits Southeast Asia. The gaur is dying out due to poaching and a loss of its native habitat. Two years later Lanza and his lab cloned another endangered species, a wild ox called the banteng. This time, instead of starting with tissue from a living animal, the team used skin cells that had been taken from an animal 23 years earlier and frozen. The project proved that if properly preserved, tissues could be stored for long periods of time and then later used to produce new animals. In some cases this might be the only way to ensure the long-term survival of endangered species.

By 2008 the group had learned enough to begin using embryonic stem cells to produce blood. Starting with one dish of embryonic cells that had taken their first steps of development, the researchers carried out a four-step procedure exposing the cell to proteins that guide the development of blood. This led to the development of 100 billion red blood cells of types A-positive, A-negative, B-positive, B-negative, and O-positive. The process was not perfect—the blood could not immediately be transfused into a patient—because the cells contained immature forms of the *globin* proteins needed to carry oxygen. But it was the first time researchers had managed to transform human embryonic stem cells into red blood cells. A strategy that is being tried in many labs is to generate stem cells, blood, and other tissues starting from adult cells. One day this, too, may work—but the method has yet to produce blood.

It is likely to take several more years before scientists manage to create red blood cells suitable for transfusion. At the moment it would be far too expensive, but Lanza and his colleagues are confident that with further work, the procedure can be developed into a method to produce blood and other tissues. "Limitations in the supply of blood can have potentially life-threatening consequences for patients with massive blood loss," Lanza said in a statement issued by ACT. "Embryonic stem cells represent a new source of cells that can be propagated and expanded indefinitely, providing a potentially inexhaustible source of red blood cells for human therapy."

him or her to fly, become invisible, or teleport to the other side of the globe.

The Dutch researcher Hugo de Vries, one of the rediscoverers of Gregor Mendel's work, proposed the hypothesis that genes might change through mutations, creating new material for evolution to work with. Early geneticists such as Thomas Morgan had to wait for mutations to happen naturally in laboratory organisms such as fruit flies; they had no way to make them happen. But from the beginning Morgan and members of his lab tried to find ways to make mutations happen more quickly with chemicals or other methods. In 1927 Hermann J. Muller (1890–1967), a former member of Morgan's lab working at Rice University in Texas, discovered that exposing flies to X-rays caused mutations. As well as providing geneticists with a new tool, the finding was a sensation and alerted doctors to the possible dangers of exposing their patients—and themselves—to radiation. It led to a Nobel Prize in physiology or medicine for Muller in 1946.

The double-helix structure of DNA discovered by James Watson and Francis Crick in 1953 provided some insights into how mutations might occur naturally. The scientists' famous paper showed that when DNA was copied, particular letters of the genetic code might be exchanged for others. The *base adenine* might be substituted for *guanine* (or vice versa), or *cytosine* might be replaced by *thymine.* But many other kinds of errors could occur. The DNA sequence might get longer or shorter through the addition or loss of chemical letters. Entire regions might be accidentally copied twice. Sequences might be cut out and pasted back in somewhere else. Errors in the way egg or sperm cells are created might produce an embryo with an extra chromosome—this is what happens in Down syndrome, in which a child inherits a third copy of chromosome 21. The most dramatic changes involve inheriting an extra copy of the entire genome, which can also begin as a defect in the creation of egg or sperm. This process is described in the next chapter.

Mutations affect an organism when they change the recipes that cells use to make RNAs and proteins. Since they happen randomly, and only a small percentage of the human genome

With their discovery of the structure of the DNA molecule, James Watson (left) and Francis Crick (right) revealed the role that DNA plays in heredity and in the chemistry of the cell. (*A. Barrington Brown, Science Photo Library*)

(less than 2 percent) encodes proteins, most mutations do not affect genes. If they occur in noncoding regions, they are usually harmless. But those that affect the chemical spelling of a gene may change the architecture and function of the protein that is made from it.

DNA is made of a four-letter chemical code that is transcribed into RNA and then *translated* into the 20-letter code of proteins (amino acids). It takes three bases of DNA to "spell" one *amino acid*. There are 64 possible ways (4^3) that the bases can be combined into three-letter "words" (called *codons*). This is more than are needed, but most amino acids can be spelled in three or four different ways. This is important to understand when thinking about the effects of a mutation. If a single base of DNA is altered, the codon that it belongs to may still spell the same amino acid. This kind of change, called a *synonymous mutation,* does not change the protein recipe.

But *nonsynonymous mutations* do lead to the substitution of an amino acid in a protein. This changes the chemistry of a protein and it may have a significant effect on its architecture and behavior. When a protein is made, the chemical attraction between its subunits cause it to fold into a precise three-dimensional shape. This puts some amino acids on the outside where they can interact with other proteins, and others on the inside where they may tie parts of the molecule together. A change on the outside may prevent a protein from binding to another molecule and prevent it from carrying out some important job. A change inside might make the protein too stiff to move, or shift parts of the molecule so that it can no longer perform chemical reactions. A *receptor* on the cell surface may no longer be able to receive an important signal, or an enzyme may cut another protein in the wrong place. Some of these effects are so serious and affect such important molecules that they are eventually fatal. Others cause organs or other body structures to be built improperly. And some have very small effects because the body has a backup system—another protein is able to step in to take over the functions of the one that has been damaged.

Some mutations prevent a protein—or large parts of it—from being made at all. One reason has to do with the fact that of the 64 codons that can be formed, two do not encode amino acids. They are called *stop codons* because they signal the end of the protein; when the translation machinery encounters one, it breaks off the process of building a molecule. If a stop codon occurs near the beginning of a molecule or in certain other places, the cell may not build a protein, or it may create only a fragment.

The duplication of a bit of a DNA sequence can have a similar effect. Adding extra bases to a gene may cause a *frameshift*—a rearrangement of codons. This would be like moving the spaces in a sentence made of three-letter words. For example, it might rewrite the sentence:

"The old men are now sad"

so that it reads:

"The ool dme nar eno wsa d"

It would be hard for a person to read a text in which this had happened, and in a gene it respells all of the codons, completely changing the recipe for a protein. Losing one or more letters of the DNA sequence has the same effect. In a long gene, a frame-shift often changes the spelling of an amino acid into a stop codon. This may break off protein synthesis too early, creating a protein that is missing important information. It may also remove a stop codon and produce a molecule that is too long, with extra information that changes its behavior. Since proteins have been fine-tuned over the course of evolution, these changes are usually harmful. On rare occasions, they may improve a molecule's functions and then spread quickly through a species through natural selection.

Most mutations are probably harmless, and a few may even provide some sort of advantage to the person who has them. An example involving the eye was discovered in 1993 by Gabriele Jordan and John Mollon, of the University of Cambridge. Human eyes are able to see a huge range of colors because their eyes contain three pigments that capture red, green, and blue light (like the "RGB" system used in many computer monitors and printers). But many insects have additional photopigments, which made Jordan and Mollon think that some humans might have extras, too. They carried out an experiment in which subjects were asked to tell whether two shades matched or not. One of their subjects, known as "Mrs. M," seemed to be a "tetrachromat"—a fourth photopigment allowed her to see an additional color between red and green.

If a mutation or another type of change in DNA happens during the creation of an egg or sperm or during fertilization,

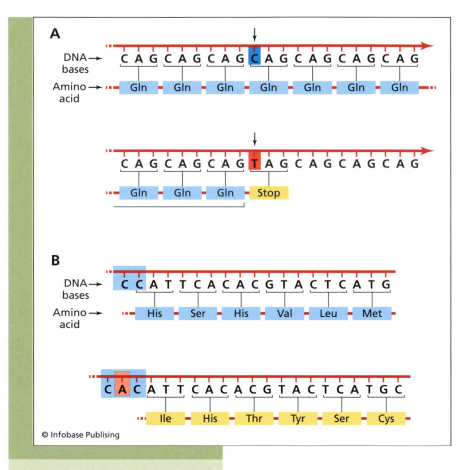

Frameshifts and mutations. Two types of mutations in genes that alter the proteins they encode: A) The switch of a single base can change the spelling of an amino acid into a stop codon, leading to a protein that is too short; B) The insertion or deletion of a single base causes a frameshift, changing the boundaries of the codons and causing the cell to use different amino acids to make the protein.

it enters a person's genome and may be passed down to his or her children. But mutations can occur throughout a person's lifetime. Such *somatic mutations* occur naturally, but they also arise through encounters with toxic substances like tar and nicotine, other factors in the environment, or radiation. These are not passed along to a person's children unless they affect egg or sperm cells, which is usually rare.

The ultraviolet rays in sunlight cause somatic mutations in skin cells that may lead to skin cancer. Doctors have documented an unusual number of tumors which begin on the right arms of Australians. The cause seems to be that as people drive (on the left side of the road in Australia), they hang their arms out the window, where they are battered by radiation from the sun.

The DNA-damaging effects of radiation have allowed scientists to turn it into a tool for fighting cancer. When cells are exposed to low doses of radiation, they suffer breaks in their DNA molecules. If the damage is not too severe, it can usually be repaired. But cancer cells are usually poor at this, and they reproduce so quickly that they pass the defects on to their offspring. This usually leads to the death of the cells. But unfortunately, in most cases the damage caused by radiation is not restricted to tumors. It does affect other cells and leads to unwanted side effects. Researchers hope that by discovering which molecules trigger the onset of the disease, they can find new therapeutic approaches that will target only defective cells.

2

The Evolution of the Human Genome

Until the second half of the 20th century human evolution was mostly a field for paleoanthropologists, who studied fossils to understand the evolution of the body, and a few brave primate researchers who spent long years among chimpanzees and apes in hopes of gaining insights into the origins of human behavior and perhaps society. The best clues about human origins came from precise measurements of fossils—usually small fragments from the bodies of adults—and comparing them to modern humans and primates. This was as unsatisfying as trying to guess how a meal was prepared by smelling the odors lingering in a kitchen afterward. With the ability to read the genetic code, scientists have obtained a direct look at the recipe book. Evolution has become a molecular science.

Understanding the evolution of the human genome may reveal what ancestors looked like and where they lived, but the main goal is not to produce a complete family tree of *Homo sapiens,* or to reconstruct skeletons for museums. Instead, scientists hope to understand how subtle changes in genes produced modern humans from earlier species by comparing the genomes of living species with each other and with DNA sequences obtained from recent fossils. They hope to learn how the body adapted in response to environmental factors such as changes in the climate or threats from new diseases. Another goal is to understand how very special human traits like cognition, language, and artistic abilities arose through variations in the genes of their ancestors and the pressures of evolution. Along the way, researchers have learned where modern humans likely originated, and how they settled the globe.

The discovery of each new hominid fossil is exciting, but fossils will never tell the complete story of human evolution, any more than picking a few random names from history books would allow someone to complete a family tree. In combination with other types of data, molecular evolutionary studies are revealing a great deal about how human biology evolved. This chapter traces that story, following the most ancient information in the genome to the most recent.

The ancient history of the genome can be read by comparing descendants of organisms that branched off from each other long ago, such as humans and bacteria. Modern evolutionary history—the past few million years—is read by comparisons to chimpanzees and apes. And ongoing evolution can be watched within a species, studying the DNA of immediate relatives, extended families, and groups of people scattered across the globe.

THE ORIGINS OF THE GENOME

The human genome is part of a living history that stretches back to the beginning of life on Earth, and a great deal of that history can be read—at least partially—by comparing the DNA sequences of humans and other organisms. Reading the information in the genome is like visiting a new country and finding artifacts from the epochs of its history. From the oldest eras a traveler would find only the remains of great fortifications, the foundations of old cities, and eroded monuments; recent eras are often represented by detailed, well-preserved information.

The oldest features that can be detected in the human genome are those shared with the most distantly related forms of life, single-celled bacteria and *archaea*. In spite of their great differences, these three main branches of life have some interesting similarities. All have membranes that protect them from

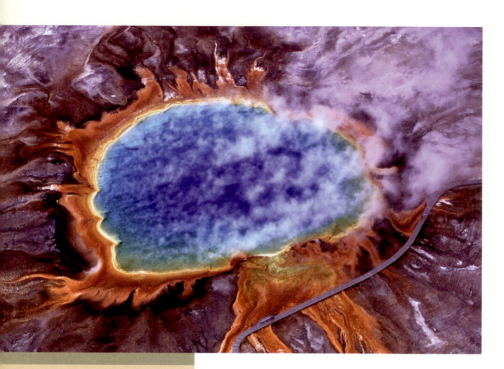

An aerial view of the Grand Prismatic Spring, a hot spring in Yellowstone National Park that is home to archaeal cells. These unicellular organisms arose early in Earth's history and were first found in extreme environments such as hot springs. (*Jim Peaco, National Park Service*)

the environment. They carry their genetic material in DNA molecules and use it to produce RNAs and proteins. Their genomes also encode the molecular tools needed to carry out these transformations. Another similarity is the fact that they obtain energy by breaking down sugar molecules. This means that their common ancestor (called LUCA, for the "last universal common ancestor") almost certainly had all of these features. Very little is known about life before the appearance of that ancestor, although biochemists are steadily learning how some of its components—such as amino acids—could have arisen from inorganic substances. The early Earth may have contained other forms of life, but nothing is known about their characteristics.

One branch of life became the *eukaryotes*—cells with a nucleus, which would later produce all fungi, plants, and animals, as well as unicellular organisms such as yeast and many parasites. There are no fossils of the first eukaryote, but researchers have learned a lot about it by comparing a huge number of existing species. All of them have DNA stored in a nucleus as well as several other membrane-enclosed compartments. They also have a sophisticated *cytoskeleton,* a system of fibers made of *tubulin* and other proteins that give cells their shapes and structure, help them migrate, and play a very important role in cell division.

Until 1991 it was widely believed that the cytoskeleton first appeared in eukaryotes. But that year Erfei Bi and Joe Lutkenhaus, two biologists at the University of Kansas Medical School, discovered a protein in bacteria that behaves like a part of the cytoskeleton. In the November 14 edition of the journal *Nature* they reported that a protein called FtsZ forms a ring-shaped structure around the "equator" of a bacterium as it prepares to divide. Creating two daughter cells requires gradually cinching this beltway until the membrane is completely pinched off. A close look at the sequence of the FtsZ gene showed that it was so similar to tubulin that one molecule had likely evolved from the other.

This finding was supported by two studies that appeared in 1998. Jan Löwe and Linda Amos of the MRC Laboratory of Molecular Biology in Cambridge, England, used X-rays to obtain a high-resolution picture of the building plan of FtsZ. At the same time, Eva Nogales's group at the University of California in Los Angeles exposed that of tubulin. Their shapes turned out to be remarkably similar, which is further evidence that they descend from the same gene.

Another feature that distinguishes eukaryotes from bacteria and archaea is the presence of compartments within the cell that are enclosed in membranes, almost like cells within the cell. At the beginning of the 20th century a few microscopists had hypothesized that this was exactly what they were—that they had originally been separate organisms. It was an interesting idea but was not very convincing due to a lack of evidence. That began to change in the 1970s when Lynn Margulis (1938–   ), a biologist at Boston University, proposed an explanation for the evolution of one of these *organelles,* called the *mitochondria.* Margulis hypothesized that it arose when an early bacteria invaded

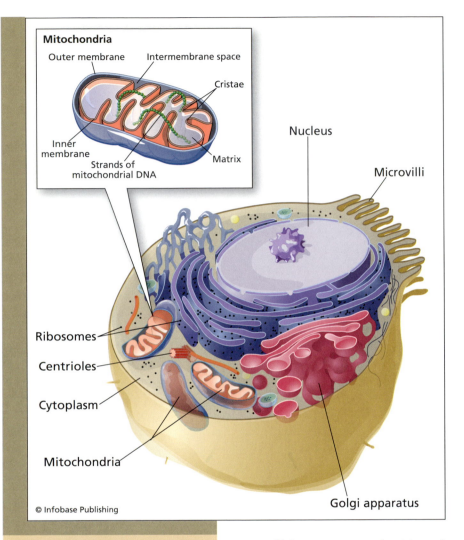

Mitochondria

Outer membrane   Intermembrane space

Cristae

Inner membrane

Strands of mitochondrial DNA

Matrix

Nucleus

Microvilli

Ribosomes

Centrioles

Cytoplasm

Mitochondria

Golgi apparatus

© Infobase Publishing

Mitochondria lie in the cell cytoplasm, outside the nucleus. A single eukaryotic cell may contain hundreds or thousands of them. Mitochondria have a complex structure, containing their own DNA, and reproducing independently of the rest of the cell. They are thought to have evolved through symbiosis, when bacteria invaded an earlier type of cell.

a cell (or was eaten by it) and stayed, forming a successful partnership with the larger cell. The two organisms eventually became completely dependent on each other. Mitochondria have adapted fully to symbiotic life. They are unable to survive on their own, and cells depend on them to provide energy by breaking down carbohydrates.

Margulis' ideas were originally regarded as so speculative that her initial paper on the topic was rejected by more than a dozen journals before it was finally printed in the *Journal of Theoretical Biology*. Since then, comparisons of the DNA of bacteria and mitochondria and other types of evidence have lent support to the idea, and most scientists now accept it. The hypothesis encouraged scientists to look for other examples of symbiosis in evolution. In the late 1980s studies by James Lake, a bioinformatician at UCLA, revealed some unexpected links between sequences of genes in archaea and eukaryotes. Ten years later this led Purificación López-García and David Moreira, two evolutionary biologists at the University of Paris-Sud, to propose that the eukaryotic nucleus was originally an archaeal cell that took up residence in a bacterium.

The idea is still being debated, but more interesting evidence has been gathered to support it. In 2001 bioinformaticist Takao Shinozawa's laboratory at Gunma University, in Maebashī, Japan, took a new approach to analyzing the relationships between genes from yeast (which are eukaryotes), archaea, and bacteria. Most earlier studies, the group pointed out, had compared different types of genes from the cells. By eliminating sequences from mitochondria (which have their own genes, presumably obtained from bacteria), they detected a clear pattern: Genes that controlled processes related to the cell nucleus were much more closely related to the molecules of archaea.

In an article published in the February 2001 issue of *Nature Cell Biology,* the scientists stated, "Our results clearly show the

Lynn Margulis in the 1960s, now professor at the University of Massachusetts, Amherst *(Lynn Margulis and Bates College)*

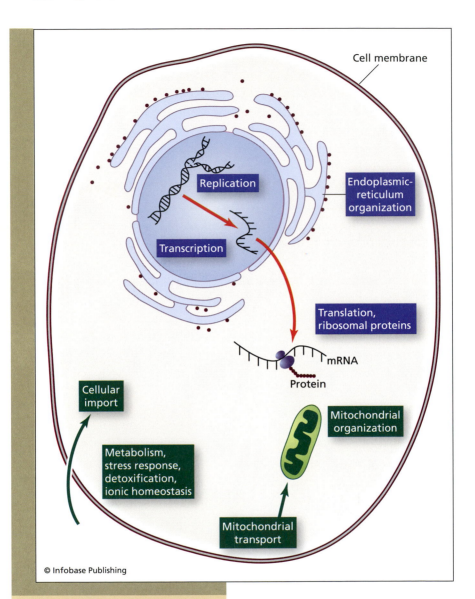

Cell membrane

Replication

Endoplasmic-
reticulum
organization

Transcription

Translation,
ribosomal proteins

mRNA

Protein

Cellular
import

Mitochondrial
organization

Metabolism,
stress response,
detoxification,
ionic homeostasis

Mitochondrial
transport

© Infobase Publishing

An analysis of the genes of modern organisms provides hints to the origins of structures within eukaryotic cells. Mitochondria and genes involved in processes such as cellular import and detoxification likely originated in bacteria (green). The nucleus and genes involved in control of the cell cycle may have come from archaea (red).

presence of a mosaic structure (derived from Archaea and Bacteria) in the eukaryotic genome." They concluded that the data strongly supported the idea that the nucleus arose from a symbiosis with archaea.

Without these events, which probably happened about 2 billion years ago, eukaryotic cells would never have evolved—nor would the plant and animal species alive on Earth today. The fact that different types of ancient cells learned to coexist has produced some of the oldest information in the human genome.

TRACES OF THE EARLY EVOLUTION OF ANIMALS

At some point, instead of going their separate ways after cell division, eukaryotic cells remained together, living in colonies. The genes that had permitted them to stick to surfaces, receive stimuli, and find food were now used in new ways, as tools to hold them together, migrate through a body, and build tissues and structures.

Most of today's animals—from insects to fish and mammals—are descendants of tiny wormlike creatures called *Urbilateria*. The original member of this family was asymmetrical from head to tail and from front to back, but symmetrical from left to right. Over time genes acting in the head region produced eyes, a brain, and a nervous system stretching down into the body. The digestive system ran from head to tail, and limbs sprouted off the torso. This basic body plan is a main theme that has been spun off in countless variations over hundreds of millions of years. The molecules responsible for animal building in the head-to-tail direction are called the *homeobox* (HOX) genes, and today the descendants of those genes are found in humans and every other animal whose body has such a plan.

HOX genes are arranged next to each other on a chromosome, like a string of words making up a sentence. As cells in the early embryos of animals divide, the HOX genes are activated one after another in cells. This happens in an unusual way, in the order in which the genes physically appear on the chromosome, like a row of falling dominoes. Another unusual thing about the HOX genes is that the order of words in this genetic sentence has been preserved in all kinds of animals for nearly a billion years. Normally over the course of evolution, genes

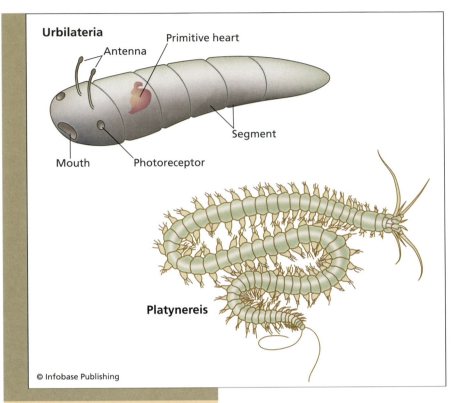

© Infobase Publishing

All animals whose bodies have left-right symmetry (including humans) arose from an ancient wormlike organism that scientists call the Urbilateria. Above: An artist's view of what it might have looked like. Below: The modern marine worm *Platynereis dumerilii* is thought to have preserved many features of this ancient organism.

lying next to each other on a chromosome are split up and become part of new sentences. In this case not only have they remained together, but HOX genes are so important that the "meaning" of the sentence has stayed virtually the same. Related genes are responsible for building similar structures throughout the animal kingdom.

No one knows what is the minimal number of genes needed to build an animal, but it is clear that complex organisms like humans need more DNA than bacteria. Many species of bacteria have about 4 million base pairs of DNA; the human genome contains approximately 3.2 billion. Lungfish, strange creatures that inhabit swamps in the Southern Hemisphere and which have both lungs and gills, have 40 times as much DNA as humans. Where did all of this extra information come from?

In the early 20th century the laboratories of geneticists Thomas Morgan and Hermann Muller (introduced in chapter 1) discovered that when cells copied their DNA, extra copies of genes sometimes appeared. Recent studies of genomes have shown that this happens nearly as often as mutations. This gradually increases the size of the genome in some species, but in others it is counterbalanced by another type of error in which genes are lost. Some of the most dramatic examples of genome reduction can be found in bacteria that have lost their ability to live outside a plant or animal; they have become completely dependent on their hosts. Like a rich guest in a hotel, they can take advantage of "room service"; the host takes care of many of their biological needs. Animals have likewise experienced genome shrinkage. In 2005 the lab of Detlev Arendt, a developmental biologist at the European Molecular Biology Laboratory in Heidelberg, Germany, showed that the genome of the fruit fly is likely to be quite a bit smaller than that of the common ancestor of insects and vertebrates.

The genomes of vertebrates, on the other hand, have expanded tremendously. In the late 1960s the geneticist Susumu Ohno (1928–2000), working at the City of Hope Medical Center of Los Angeles, made the claim that gene duplications have been the most important factor in the evolution of genomes. Initially a gene duplication usually produces identical copies of a molecule, leading an organism to make twice as much of a particular protein. But over time the molecules undergo different types of mutations that can give them new functions.

The human genome is full of examples, such as the presence of nine closely related globin genes. Multiple copies of the genes are also found in birds. A bird called the Rüppell's griffon builds four different types of *hemoglobin,* giving it the amazing ability to fly at altitudes of six miles (10 km) above sea level. Other gene duplications have greatly enhanced the immune system of mammals. The human genome contains more than 150 immunoglobulin genes, which can be linked together in an almost infinite number of patterns to make antibodies, one of the topics of chapter 4. Gene duplications gave primates a system of vision based on three colors, and they have also greatly enhanced mammals' sense of smell by spinning off new molecules in sensory cells in the nose. The number of such *olfactory receptor genes* has gone down again as humans have lost many of these genes, compared to chimpanzees. Interestingly, some subtypes of smell genes have been maintained in humans. Yoav Gilad and Svante Pääbo of Leipzig, Germany—two specialists in the area of fossil DNA—think the fact that humans cook their food might explain both the loss of some types and the preservation of others.

Vertebrates often have three or four copies of genes that are found only once in other branches of life. This led Ohno to a much more dramatic idea: He proposed that the entire genome had been duplicated in an ancient organism at least once, and possibly twice. This is not hard to imagine. When cells divide, the entire genome is copied, and then the two sets are equally divided into separate compartments. If something goes wrong with the compartmentalization machinery, one cell could be left with two copies. This can happen with the entire genome or single chromosomes; such "sorting problems" lie at the root of diseases like Down syndrome, where a fertilized egg receives three copies of chromosome 21. Whole genome duplications are known to have happened in plants such as wheat and rice (possibly as a result of humans selecting plants that provide more food) and in species of fish.

Ohno's hypothesis suggests that the two duplications happened in vertebrates within a short time, just as they started off on a new evolutionary branch. It might even have been the cause of their divergence by creating a huge number of new genes that could undergo mutations and acquire new functions.

(opposite page) Genomes have expanded enormously since the first cells appeared on Earth through natural errors that happened as cells divided. Susumu Ohno proposed that early vertebrates experienced two duplications of the entire genome, giving them four copies of each chromosome pair. Single genes have also undergone multiple duplications.

The hypothesis has not yet been proven because some of the "tracks" have been erased. Humans and other vertebrates have four copies of some genes, but only one, two, or three of many others. Most scientists are convinced that one genome duplication happened; there is more debate about the second. Rather

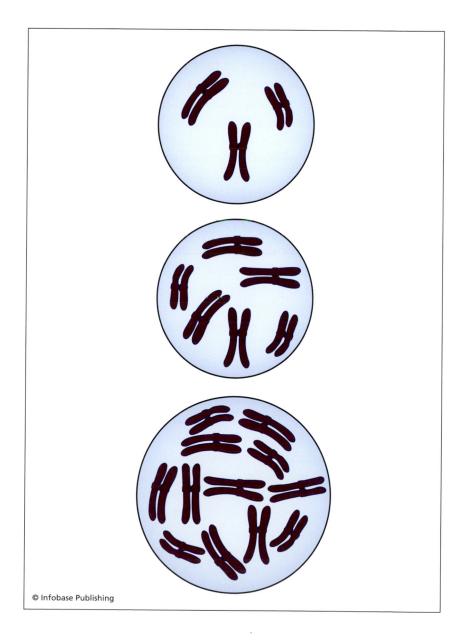

than a single event, there may have been several smaller ones in which large portions of DNA were copied. It may be possible to prove or disprove the hypothesis when complete genomes are available for a wider range of animals.

As researchers completed the human genome, they were startled to discover that a huge amount of it seemed to be made of small bits of information—usually about 300 bases long—repeated over and over. These elements are called *Alu sequences,* or Alu repeats. (The name comes from the fact that the sequence was first identified in a bacteria called *Arthrobacter luteus.*) Scientists estimate that there may be 1 million copies in the genome—they may make up more than 10 percent of the entire human sequence. They are scattered everywhere, almost as if someone copied a sequence hundreds of thousands of times and pasted it in again at random places. In fact, this scenario is probably not far from what actually happened.

The first Alu sequence was likely brought into human cells by a virus that behaved like HIV, which causes AIDS. HIV is a retrovirus, which means that it reproduces by inserting its genetic material into the DNA of the host. Like most viruses, it carries only what it needs, taking advantage of molecules in the cells it infects to survive and reproduce. For instance, instead of carrying along molecules to transcribe its own genetic information (which is made of RNA), HIV borrows them from the cell.

Upon entry into the cell the virus is taken apart, and its RNA and some of its proteins enter the nucleus. One of these molecules is called a *reverse transcriptase*—a tool that carries out transcription in reverse. Instead of reading DNA and making a single-stranded RNA, it reads RNA and assembles a single strand of DNA. Another protein tool stitches this strand into the cell's genome. Once there, it is treated like any other human gene; the cell transcribes it into RNA that is then transported to the *cytoplasm.* Some of the RNA is translated into proteins that the virus needs. These are then collected with RNAs and packed into membranes to make new copies of the virus.

Reverse transcriptases were independently discovered by two virologists in 1970: Howard Temin of the University of Wisconsin–Madison and David Baltimore at the Massachusetts

Institute of Technology. The discovery was dramatic because it obviously contradicted the "central dogma's" position that information could only flow from DNA to RNA and not vice versa. It also helped explain the behavior of viruses that had been linked to the development of cancer. Both Temin and Baltimore had begun their work on viruses in the laboratory of Renato Dulbecco at the California Institute of Technology (Dulbecco then moved to the Salk Institute), and for their discoveries the three men were awarded the 1975 Nobel Prize in physiology or medicine.

If DNA from a virus makes its way into egg or sperm cells, it becomes part of an animal's genome. In following generations it may continue to behave the same way even in the absence of a virus. The genes are transcribed into RNA, which is then reverse transcribed into DNA and reinserted somewhere else in the genome. Such molecules are called *transposons,* or "jumping" genes, in contrast to normal genes, which usually occupy fixed locations. They were discovered by geneticist Barbara McClintock (1902–92) in the 1930s and 1940s while working at the University of Missouri and the Carnegie Institution of Washington's Genetics Research Unit at Cold Spring Harbor in New York. McClintock was an excellent microscopist and bold thinker who discovered transposons while working on maize. Jumping genes seemed to be the only way to explain how the colored patterns of corn kernels were passed from generation to generation. The idea was so radical that very few of her contemporaries understood or believed her findings. But the discovery of reverse transcription opened the door to the idea that cells might contain their own mechanisms for converting RNA into DNA, and McClintock's work was finally appreciated—half a century after she first proposed the idea. She was awarded the Nobel Prize in physiology or medicine in 1983.

The Alu sequence and jumping genes have been a major factor in evolution by increasing the size of genomes, giving evolution fresh new material to work with. It is estimated that about 100 human genes have been "domesticated" in this way. The gene Harbi1, for example, which probably plays a role in repairing DNA, was acquired from a transposon called Harbinger.

But such domestications are relatively rare. Transposons are usually dangerous because inserting DNA sequences into random places in the genome eventually causes problems. Sometimes they land in the middle of another gene, which nearly always destroys it. They are so dangerous that natural selection has inactivated nearly all of the known transposons in vertebrates, except for the domesticated genes, which have taken on other jobs and no longer jump.

MOLECULAR CLOCKS

Until researchers learned to sequence genes and proteins, they had no idea how quickly mutations changed the genetic code. The French geneticist François Jacob (1920–), one of the pioneers of modern biology, said, "When I started in biology in the 1950s, the idea was that the molecules from one organism were very different from the molecules from another organism. For instance, cows had cow molecules and goats had goat molecules and snakes had snake molecules, and it was because they were made of cow molecules that a cow was a cow." Instead, scientists discovered that the pace of mutations had not obscured the history of evolution. The related genes of cows, snakes, and humans could be identified and compared. With enough data, it was possible to distinguish the changes from the original and determine the DNA sequence of the ancestor's gene.

In 1962 chemists Emile Zuckerlandl (1922–) and Linus Pauling (1901–94) of the California Institute of Technology realized that these relationships could be turned into another kind of tool to study evolution. They were analyzing the sequences of hemoglobin proteins in different species. In some cases the differences were extreme, in others very small. The number of changes in the sequence was not random; it roughly corresponded to the amount of time that had elapsed since the two species had diverged, as best could be determined by the fossil record. If changes happened at a steady rate, Zuckerlandl and Pauling reasoned, then DNA sequences could be used to time evolutionary events. The idea was picked up by Allan Wilson (1934–91), a native of New Zealand who had come to the University of California at Berkeley to do his Ph.D. and remained there for his entire career. Wilson gave the approach a name: the *molecular clock.* Counting the differences between two species would reveal the date at which they had diverged from a common ancestor. He went on to use the method to make several pioneering discoveries about human evolution, discussed later in the chapter ("Mitochondrial Eve and Y-Chromosomal Adam").

This basic concept has been helpful in estimating times of important evolutionary events. Real clocks are useful because they measure time at an even pace, but mutations do not really behave that way. The mutations that affect evolution usually occur during the creation of egg and sperm cells or their fusion during fertilization. This means that the clock ticks faster in species that reproduce quickly. To take an extreme example, 100,000 years is about the equivalent of 6,000 or 7,000 generations for humans. The comparable ancestor of a fruit fly—grandfather to the 7,000th degree—lived only 230 years ago. If its rate of reproduction remained the same, it would undergo 3 million generations in 100,000 years. This has to be taken into account when estimating the time at which species such as humans and flies diverged. At the moment that happened their clocks were synchronized, but gradually they began to run at vastly different rates.

Some organisms may be better at detecting errors in DNA and repairing them than others, which also has to be taken into account in calibrating the clock. Another adjustment has to be made depending on the part of the genome at which a researcher is looking. Genes usually change at a much slower rate than DNA sequences that do not encode proteins, because they often cause problems for an organism. If they prevent it from surviving and reproducing, they will not be passed along to future generations. Noncoding sequences are far less likely to make a difference, so these parts of the genome serve as the best clocks for recent events. On the other hand, they change so quickly that they cannot be used to make guesses about events that happened too long ago. At that point researchers turn to genes.

Every application of the molecular clock depends on assumptions about factors such as these. In some cases those assumptions can be tested by comparing dates given by the clock to the fossil record. Sometimes this cannot be done very accurately, especially for ancient events. In an article published in the online journal *Genome Biology* in 2001, biologist Gregory Wray of Duke University, in Durham, North Carolina, points out that there is often a discrepancy between hypothetical dates proposed by clock models and fossils. He writes: "On the basis of fossil evidence, the great divide between prokaryotes and eukaryotes occurred about 1.4 billion years ago (Ga); estimates from sequence data suggest earlier divergence times of 2.1 Ga for the split between archaebacteria and eukaryotes and over 3 Ga for the split between eubacteria and eukaryotes."

As a rule, Wray says, molecular biologists tend to give older dates for events than paleontologists. That is understandable because neither method is perfect, there are gaps in the fossil record, and perhaps most importantly, they measure different things. According to Wray, "Sequence differences reflect the time since two *taxa* last shared a common ancestor (their divergence time), whereas fossils reflect the appearance of anatomical structures that define a specific group (its origin). The two events may be widely separated in time: early members of a group can be quite different in anatomy, habitat, and size from later, more familiar members." In other words, DNA sequence data identifies the great-grandparents (many times removed) of fossil children, and there is no telling how many generations really separate them. Scientists are most likely to find a fossil once the species it belongs to has formed a large population, which may be a long time after it first evolves.

A recently discovered fossil from Western Africa (see "The Search for the Earliest Hominid") presents an unusual case in which the situation might be reversed: A skull found in Chad has characteristics of both chimpanzees and humans and its discoverers believe it might represent the common ancestor of both, or at least a close relative of that ancestor. Chimpanzees are the closest living relatives of *Homo sapiens,* and the ages at which the two species reproduce are still fairly close, so if ever the clock method should be accurate, it seems like this should be the case. Current estimates place the date of divergence at somewhere between 4.5 and 6.5 million years ago (MYA). The fossil that was found in 2001 has been dated to about seven MYA.

The dates are close but do not match, which reveals the main problem in trying to reconcile the two types of data. DNA sequences cannot currently be obtained from samples older than about 100,000 years; it may never be possible to obtain them. But this information might be the only way to definitively prove that one organism was the parent of another, rather than an aunt or uncle, a cousin, or some more distant relative.

COMPARING THE CHIMPANZEE AND HUMAN GENOMES

How much of human nature is truly unique to *Homo sapiens* and how much is shared with apes, vertebrates, and other animals? An important step to finding an answer came in 2005 when, for the first time, scientists were able to contrast the genome of human beings with that of their closest evolutionary relative, the chimpanzee. The completion of the chimp genome was announced in the September 1, 2005, issue of the journal *Nature* by the Chimpanzee Sequencing and Analysis Consortium, a group of 67 researchers from 24 international laboratories.

They found 40 million changes that distinguish the genomes of chimps and humans. That seems like a huge number until one considers that the total human sequence consists of approximately 3.2 billion base pairs. Most of the differences are small—changes in single letters of the code. Entire blocks of the two genomes can be directly compared to each other because they contain the same information, in the same order, and those regions are 99 percent identical. Comparing the rest takes ingenuity because of sequences that have been duplicated, lost, or rearranged. Even taking those into consideration, the genomes are 96 percent identical overall. To get a feeling for what this means, consider that any given chimp and human differ from each other

The Search for the Earliest Hominid

In 2001 an expedition headed by paleontologist Michel Brunet, of the University of Poitiers, France, discovered a remarkable fossil in the Djurab Desert in Chad, Africa. Brunet and a team of 40 scientists had been working the site for several years, uncovering thousands of remains of ancient species. Many experts believed the region to be an unlikely place to discover *hominid* fossils—it was far to the west of the Great Rift Valley, the area that had been worked for decades by the family of Louis Leakey (1903–72). Dozens of major finds in Olduvai Gorge and other parts of the valley had helped convince paleontologists that Eastern Africa was the cradle of humanity, the place where hominids had begun to walk upright and taken crucial steps on the evolutionary path to modern humans. But in 1995, at the new site, Brunet had unearthed the remains of a new species of Australopithecines, determined to be 3.5 million years old. When the find was followed by many more, paleontologists began to shift their attention to the west.

Even so, the 2001 discovery was completely unexpected. During a walk across the site, undergraduate student Ahounta Djimdoumalbaye from the University of N'Djamena in Chad uncovered a nearly complete cranium, teeth, and part of the lower jaw of a type of primate that had never been seen before. It had a braincase like that of a modern chimpanzee, but the researchers were excited by the position of its *foramen magnum,* a hole where the spine enters the base of the skull. The gap was farther underneath the skull than in apes, which is usually interpreted to mean that the creature walked upright—it was a hominid. This meant that the find might represent an ancestor of humans, or a close relative of an ancestor. That was especially interesting in light of the fossil's age. A careful study of the sediments in which it was found showed that it was at least 6 million years old. Further work has revealed that the animal probably lived about 7 million years ago. If it was an upright-walking hominid, it was the earliest that had ever been found.

Brunet and his colleagues christened the new species *Sahelanthropus tchadensis* and nicknamed it "Toumaï," a local word meaning "hope of life." In an article announcing the find in the July 11, 2002, issue of *Nature,* Brunet stated that features of the teeth and face were similar to later hominids, including the ancestors of humans. Other features were reminiscent of chimpanzees. "The observed mosaic of primitive and derived characters evident in *Sahelanthropus* indicates its phylogenetic position as a hominid close to the last common ancestor of humans and chimpanzees," he wrote. "Given the biochronological age of *Sahelanthropus*, the divergence of the chimpanzee and human lineages must have occurred before 6 Myr, which is earlier than suggested by some authors." The last remark was an understatement; most scientists believed that the split had occurred at least a million years later. This hypothesis was partly based on comparisons of human and chimpanzee genes, which will be discussed later.

Over the next few years Brunet's interpretation of the find was widely debated by scientists. In 2006 Milford Wolpoff, a paleoanthropologist at the University of Michigan, pulled together the evidence in an article entitled "An Ape or *the* Ape: Is the Toumaï Cranium TM 266 a Hominid?" printed in the journal *PaleoAnthropology*. The article focused on the teeth and whether features of the cranium really indicated that Toumaï walked upright. Wolpoff believed that Brunet had used an outdated method to interpret the teeth, and also that the structure of the foramen magnum had been misunderstood.

(continues)

(continued)

Wolpoff concluded, "*Sahelanthropus* was an ape living in an environment later abandoned by apes but subsequently inhabited by australopithecine species. . . . Yet, it is a highly significant discovery . . . perhaps mostly because of the insight it might give for understanding the ancestral condition before the hominid-chimpanzee split." Gaps in early human evolution continue to be filled. In October 2009 *Science* reported the discovery of a 4.4-million-year-old hominid skeleton in Ethiopia, named *Ardipithecus ramidus*. This new member of the human family tree is 700,000 years older than the *Australopithecus* fossil "Lucy." Scientists believe it is a descendant of the common human-chimpanzee ancestor and may stand in a direct line to humans: Its features are more humanlike than chimp and suggest that our ancestors may have walked upright earlier than previously believed.

only about 10 times more—genetically speaking—than any two randomly chosen humans. For another comparison, it makes humans and chimpanzees 10 times more similar than mice are to rats. The mouse genome, which was completed in late 2002, was used as a control to determine whether particular changes were happening more quickly in chimps or humans.

A number like 40 million does not necessarily say much about the physical and behavioral differences between two types of animals. A mastiff and a Pekingese look more different from each other than a human and chimp, most people would say, but both breeds of dogs are the descendants of an ancient wolf, and there are six times fewer differences between their genomes. Other species can have the same amount of difference as humans and chimps yet resemble each other closely—that is the case for two types of mice called *Mus musculus* (the "house mouse") and *Mus spretus* (the "Algerian mouse").

The 1 percent difference between *Homo sapiens* and its closest relative is not spread equally across the genome. Protein-encoding regions differ by only from one to two amino acids per molecule. About 30 percent of the proteins encoded by their genes are completely identical. But 4.4 percent of the species' genes are changing much more rapidly than the norm. Those are especially interesting because they are likely to be most responsible for the differences.

Several of these molecules have something to do with the immune system and diseases. One is glycophorin C; it helps determine whether the parasite that causes malaria can infect red blood cells. It is little wonder that humans have evolved defenses against the disease. Malaria has plagued humans for more than 10,000 years; in the late 19th century its impact was so terrible that the Italian government declared that it posed the greatest threat the country had ever faced. Over the past century it has been almost eliminated in the developed world, thanks to efforts to control the mosquitoes that carry the parasite, but it is still a huge problem in many tropical climates. When malaria attacks and kills children, it prevents them from passing along their genes, so any mutations that offer immunity are likely to spread quickly through the human population. This explains the presence of certain mutations that affect human hemoglobin. Even though these mutations cause another deadly disease—sickle-cell anemia—they offer protection from malaria. So human populations living in areas affected by malaria have unusually high rates of sickle-cell anemia.

Other diseases have left their mark on the genome as well. There have also been rapid changes in a molecule called granulysin, which helps fight off the tuberculosis bacterium and other parasites.

Evolution depends on random changes in DNA sequences, many of which are harmful. An example is the gene capsase-12, found in the common ancestor of mice and primates. In the brain it launches a self-destruct program in neurons that do not function properly. Capsase-12 seems to limit the damage of protein deposits that accumulate between brain cells as an animal ages. Mutations have made the human version of the gene defective, which may partially explain why people develop Alzheimer's disease while mice and chimps do not. Chimps, on the other hand, have lost three genes that regulate the body's response to infections, which helps to explain differences between the immune systems of the species. Other genes that have evolved rapidly are molecules related to hearing—which have probably contributed to the development of language—and sensitivity to pain. But some of the most interesting changes involve the genetics of sex. Chapter 1 (see "Cells, Chromosomes, and Sex") describes how proteins encoded on the male Y chromosome influence the development of sexual characteristics.

The earliest sexually reproducing animals were probably *hermaphrodites,* which had the reproductive organs of both sexes, such as today's snails and earthworms. The males and females of these species have identical genes. Susumu Ohno, who proposed the hypothesis that the vertebrate genome had undergone complete duplications (see "Traces of the Early Evolution of Animals" earlier in the chapter), proposed that the Y chromosome evolved from a second X chromosome that became damaged, leaving only a fragment.

Analyses by David Page, a geneticist at the Whitehead Institute in Cambridge, Massachusetts, and an expert in Y chromosome evolution, suggest that the original fragment contained about 1,000 copies of genes on the X chromosome. But over time they have steadily been lost. Only 16 of the copies remain active in humans; chimpanzees have these and five more. At the same time, the Y chromosome has acquired new genes that have jumped there from other parts of the genome. Page reports that this male-specific region now makes up 95 percent of the chromosome and encodes 27 known proteins.

Not surprisingly, most of them have to do with the development of sexual characteristics and male-specific biology. Many, for example, are used only in the testes and are responsible for producing sperm cells. Once two sexes had evolved, females no longer needed these genes and they gradually disappeared from other parts of the genome. The Y chromosome served as a sort of "parking space" where such male-specific genes could be protected, since they are only passed from father to son. So gradually males and females have experienced slightly different types of evolution. Most of this has involved changes in genes only found on the Y chromosome, because males still have one X. They inherit it from their mothers, so the genes located on it cannot really undergo changes that are specific to females.

What has happened in females in many species, however, is the development of special mechanisms to carry out "dosage control." Genes on the X chromosome are important to both sexes, but there is a problem: Females have two copies of the chromosome and its genes, whereas males have only one (except for the 16 active duplicates still found on the Y). For the rest, two copies would normally cause female cells to double the amounts of proteins made from these genes, but the two sexes nearly always need equal amounts. Various species solve this problem in different ways; in mammals, females shut down the copies on one of their X chromosomes.

Page's work has focused on the Y chromosome. In some animals the loss of Y genes has become extreme—in the kangaroo, for example, only one gene is left. Grasshoppers have no Y at all; the difference between the sexes lies in the fact that females have two copies of the X chromosome and males only one. Could the same thing happen to humans? At first glance it seems possible—maybe even likely. Since mutations can strike anywhere at any time, it is always dangerous to have only one copy of a gene—as evidenced by many genetic diseases for which men are at higher risk than women. The loss of the chromosome would not necessarily eliminate maleness, just as it has not for the kangaroo, because of the other ways that genomes can produce differences between the sexes.

But the Y chromosome will not necessarily be lost. In 2003 Page's laboratory discovered that human cells have a mechanism to protect the X-related genes on the Y chromosome. First, there are two copies of them on the chromosome. During cell division they line up alongside each other, which allows the cell to detect mistakes and correct them. But when Page examined data produced by the chimpanzee genome project, he discovered that the animal does not have similar protection mechanisms. It would be interesting to see whether chimps lose the Y chromosome, or whether it will be protected through the evolution of another mechanism. However, today's scientists will never know. It took about 6 million years for five of the chimpanzee genes to vanish; losing the rest may take much longer—supposing that chimpanzees manage to survive at all. Humans have already destroyed most of their natural habitats.

THE EVOLUTION OF THE HUMAN BRAIN

Toward the end of his life Francis Crick, famous for his role in the discovery of the structure of DNA, wrote, "Consciousness is the major unsolved problem in biology." For over two decades he had been "Thinking about the Brain" (the title of an article he wrote for *Scientific American* magazine in 1979), and with a colleague named Christof Koch (1956–) he began a quest to understand how the biology of the brain could produce this unique phenomenon. Other animals may have a sort of consciousness, but only humans are able to discuss it with each other, study it, and write about it. Somehow consciousness is linked to the physical brain, and somehow evolution produced it, along with the other unique mental abilities of humans. One of the goals of evolutionary research is to understand how changes in the genome produced a brain that could develop consciousness and invent tools, language, the arts, and even artificial intelligence. The physical basis of these abilities, too, must be encoded in the part of human DNA that differs from chimpanzees and other animals. This chapter presents the genetic side of those abilities; chapter 5 explores the impact of evolution on behavior and society in more depth.

The first evolutionary studies of the brain were limited to comparing its overall anatomy to that of living animals or fossil hominids. Since soft tissue is not preserved along with bones, ancient brains could not be directly examined; scientists were limited to studying the shape and volume of fossil craniums and patterns on the inside of skulls. Those features could be compared with the brains of modern primates. They showed that evolution had given the average modern human an unusually

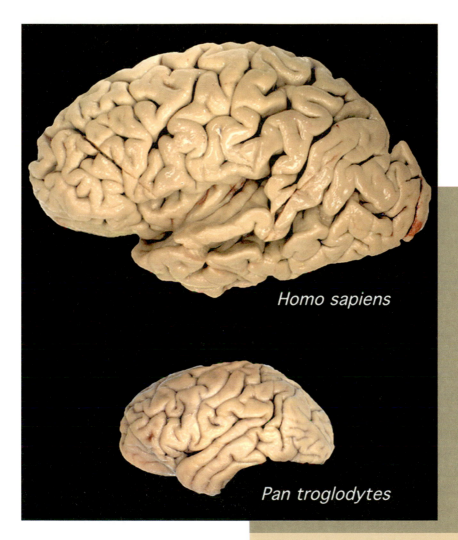

Homo sapiens

Pan troglodytes

A comparison of human and chimp brains. Until very recently, studies of the evolution of the brain were mostly limited to comparing features of their basic anatomy. Today scientists investigate the genes that contribute to differences in their size and development. *(Todd Preuss; Yerkes Primate Research Center)*

large brain compared to other primates and species. This development happened quickly. Comparisons to chimps and apes revealed that the brain of the last common ancestor was only 300 to 400 grams in size. The modern human brain generally ranges from 1.2 to 1.8 kilograms, and studies of fossils have shown that most of this growth has taken place in the past 2 million years.

Size is relative, of course, but even compared to the largest animals on Earth, *Homo sapiens* has a huge brain.

The spurt in size is due to changes in genes, but which ones? Recent genetic research carried out by Bruce Lahn of the Howard Hughes Medical Institute at the University of Chicago has pinpointed some likely candidates. In a paper that appeared in the journal *Cell* in 2004, Lahn and his colleagues examined 214 genes known to play a role in the development and function of the brain. They looked at two types of primates (humans and macaque monkeys) and two rodents (mice and rats). This allowed them to compare the rates at which brain-related genes had been evolving over the past 80 million years—the time at which the common ancestor of all four species lived. The study showed that the genes were evolving more quickly in both humans and macaques than in rodents. In the line leading to humans it had accelerated even more. This effect was the strongest for genes that help shape the brain's structure—a clear sign, Lahn says, that strong evolutionary pressure has been at work.

It does not necessarily require very many genes to increase the size of part of an organism. Mutations in only five genes, for example, were needed for Native Americans to transform a tiny plant called teosinte into the tall cornstalks that feed the world today. Lahn's study showed that something much different had happened to the human brain. A large number of genes had undergone accelerated changes.

This does not mean that the brain has violated any evolutionary "speed limits"—particular features in other species have undergone equally rapid changes. Immune system genes evolve very quickly because diseases are a major force in natural selection. They can change the genetic profile of a species in a very short time by killing huge numbers of plants and animals, leaving only individuals with certain forms of genes to reproduce. A virus might cut a swath through a population, eliminating animals that are otherwise well adapted to their environments and leaving only a few unusual members of the species behind. Mice and other rodents depend heavily on smell, so evolution has worked hard on genes related to this sense.

Lahn and several others, including neurologist Christopher Walsh of Harvard Medical School and Geoffrey Woods, a clinical geneticist at the University of Leeds in the United Kingdom, believe that a few genes have played a particularly important role in increasing brain size. Their conclusions are based on studies of genes that, when mutated, cause humans and animals to develop unusually small brains. A rare genetic problem in humans, for example, causes microcephaly—a condition in which people are born with very small brains, sloping foreheads, and narrow faces. They are often severely mentally disabled.

By 2002 studies in humans and mice had revealed defects in several molecules—including microcephalin and a gene called ASPM—that play a role in microcephaly. If a mutation leads to a very small brain, there is a good chance that the healthy version of the gene has contributed to the development of a large one. Both of these molecules have evolved unusually quickly since the split between humans and chimps. Recent experiments with mice carried out by Walsh's lab point to another gene, Nde1, which is probably involved. "The loss of Nde1 causes neurons to mature prematurely," Walsh says. "That stops them dividing so the mice end up with small brains."

"Systems neurobiologist" Jon Kaas of Vanderbilt University, in Nashville, Tennessee, points out that size alone does not account for the special mental abilities of humans. Parts of the brain helped species not because they were bigger, but because they were better at doing certain things. Kaas hopes to learn what those things were by studying brain structure. The ability to hear and produce a wider range of sounds, for example, was surely important in the development of language. Once people began to communicate, a "feedback loop" likely began, pushing ears to become even sharper and the voice to articulate more clearly.

Fossil brains would probably not allow Kaas to say anything specific about prehistoric languages. But as scientists learn to link particular regions of the brain to specific skills, the shape and contour of the inner skull might reveal what abilities were being sharpened through natural selection at various times. Discovering a rapid spurt in the brain's hearing centers

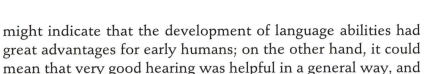

might indicate that the development of language abilities had great advantages for early humans; on the other hand, it could mean that very good hearing was helpful in a general way, and language evolved as a lucky side effect.

The true answer is probably a combination of both. Neither genes nor the brain evolved "in order to" do language—before the evolution of primitive speech, humans had to have structures in the brain and a vocal apparatus capable of doing it. Still, once those structures were in place, they would be slightly different from person to person. Those differences might make some people better at language, and then natural selection could push them in a certain direction. This kind of thinking is speculative, because it is impossible to go back in time and reconstruct all of the factors that helped some individuals reproduce more than others. But it can still help researchers understand how a population of one type of primate gradually changed into another.

In 2002 the laboratory of Wolfgang Enard at the Max Planck Institute for Evolutionary Anthropology in Leipzig, Germany, found a gene that may have played a key role in the evolution of human language. Several facts made the molecule, called FOXP2, stand out. First, it seems to help give humans precise control over their facial muscles, which is important for speech. Secondly, the researchers compared the versions of the gene found in humans, chimpanzees, and several other primates and found two changes that are unique to humans. The mutations seem to have happened at the same time as the first modern humans appeared. The most compelling evidence for FOXP2's role in the development of language, however, comes from studies of three generations of a family that has a defective form of the gene. Family members who inherited the mutations also suffer from severe problems in their ability to speak and use grammar.

In a review in the March 2008 issue of the *Brain Research Bulletin,* Jon Kaas brought together a wide range of information that had been gained from studies of primate and hominid fossils as well as living animals. Studies of casts made from the inner surfaces of fossil brains show that primates evolved from early mammals that had small brains and little *neocortex*—the "wrapping" at the top and front of the brain where a lot of the most dramatic growth has occurred. Paleontologist Robert Holloway of Columbia University, in New York City, a pioneer in the field, makes the casts by pouring vulcanized rubber into fossil skulls, waiting for it to set, and then removing the rubber through the hole at the base of the skull. Although this reveals only surface features of the brain, Holloway believes that the approach may allow him to pinpoint the stage in evolution at which language structures evolved.

Comparing the fossils with modern animals whose skulls have a similar size and form suggests that "the small neocortex of early mammals was divided into roughly 20–25 cortical areas," Kaas reports. The evolution of primates brought even more: at least 10 regions devoted to vision alone, "and somatosensory areas with expanded representations of the forepaw." The changes allowed primates to absorb a vast amount of visual information, and they were becoming very sensitive in their front paws—later to become hands.

Cortical areas, Kaas explains, are "functionally distinct processing organs." They handle different types of sensory input from different parts of the body. In living organisms, brain researchers define modules by studying how they are wired up to various parts of the body and other regions of the brain. Anatomical studies of nerves provide part of this information. Watching interactions between modules can be done with imaging techniques such as magnetic resonance imaging (MRI) that study changing patterns of fluids through the brain as it carries out various tasks.

Scientists can make a good guess about a module's functions by looking at the architecture that links it to the rest of the body. A module for sensing pain, for example, would have to be connected to nerves in the parts of the body that can feel it and then regions of the brain that govern a response. When a person hits his finger with a hammer, he feels pain because nerve cells just under the skin extend long, cablelike *axons* to a region at the border of the spine. There the impulses from those nerves are received by the *dendrites* of other neurons called nociceptors, which pass them to particular regions of the brain—pain modules.

As the neocortex increased in size and became more complex, it was able to handle more types of sensory information and a greater volume of it. Kaas writes, "As larger brains evolved in early apes and in our hominin ancestors, the number of cortical areas increased to reach an estimated 200 or so in present day humans, and hemispheric specializations emerged."

Having a large number of modules devoted to the same sense allows sophisticated parallel processing. But there may be a limit on how many types of input can be received and how much parallel processing can be handled by a brain with a particular structure. If the input overloads the brain, it might need a new mechanism to cope, and this is where Christof Koch sees an important evolutionary function for consciousness.

Koch points out that a great deal of human behavior is carried out without consciousness. "I do things—complicated actions like driving, talking, going to the gym, cooking—automatically, without thinking about them. . . . Science has provided credible evidence for an entire menagerie of specialized sensory-motor processes, what I call *zombie agents,* that carry out routine missions in the absence of any direct conscious sensation or control. You can become conscious of the action of a zombie agent, but usually only after the fact, through internal or external feedback."

Zombie agents might allow an animal to be able to do everything it needs to survive—up to a point. If too many agents are active at the same time, the brain might suffer an overload that prevents the animal from reacting properly to stimuli. One way to solve the problem would be for an organism to develop the ability to pick and choose between inputs and switch some of them off. Such a centralized decision-making mechanism is one way to describe consciousness.

Do only humans have it? Probably not, Koch says. "It is plausible that some species of animals—mammals, in particular—possess some, but not necessarily all, of the features of consciousness. . . . I assume that these animals have feelings, have subjective states. To believe otherwise is presumptuous and flies in the face of all experimental evidence for the continuity of behaviors between animals and humans."

Even so, there are differences between human and animal consciousness, and molecular biologists would like to "corner" them in the parts of human brain anatomy and the genome that distinguish humans from every other living creature. Anatomically the most unusual part of the brain is the neocortex, and studies show that it has a networking function—processing and linking input from various senses. This fits well with Koch's concept of consciousness as a coordinator of stimuli, and it may also help explain why human consciousness is unique.

MITOCHONDRIAL EVE AND Y-CHROMOSOMAL ADAM

The differences between the DNA of two species show what makes each unique. The similarities, on the other hand, tell an equally interesting story. Finding nearly identical spellings of a gene in two species points to the way it was spelled in their last common ancestor. So even if researchers have not yet found a fossil of the primate that spawned both chimps and humans—even if they never find it—they can say a lot about its genome. The same basic approach can be used to find a common ancestor for two people. What if it were used to look at the DNA of everyone in a country, or even everyone on Earth? In the late 1980s scientists began an intensive effort to collect and compare samples from people across the globe. Today this information has led to the identification of two ancestors, nicknamed "Eve" and "Adam," who belong to the family tree of every human being alive today.

The names were borrowed from the creation story in the Bible, but scientists are not using them to refer to the first *Homo sapiens.* Nor were they a real pair. They never met; they lived at least tens of thousands of years apart. This situation might seem odd, and is explained in detail below, but it has to do with the way evolution works, how humans reproduce, and the type of DNA that scientists used in the studies. Another fascinating thing about this work is that it draws on all of the themes of this chapter: the evolution of the male Y chromosome, the idea of turning back the molecular clock to find out when an ancestor lived, and even the symbiosis between ancient bacteria and eukaryotes. "Mitochondrial Eve" and "Y-chromosomal Adam" are the result of weaving these topics together.

The search for these two individuals—and they did exist, even if not in the way portrayed in the book of Genesis—began in earnest in 1987, although it followed decades of research on human blood aimed at understanding the spread of human populations throughout the globe. Years before scientists found a way to sequence DNA, they began using blood groups and types (which are inherited in Mendelian patterns) to try to build family trees for humanity. Many of the researchers expected that this method would explain the differences between races and other historical groups. But the resulting picture was confusing. As far as blood types are concerned, race does not exist. From the point of view of the genes that control blood types, populations across the world look very much the same.

Starting in the 1960s, Allan Wilson began using his molecular clock approach to address questions about human history. He had already applied the method to primates, coming to the conclusion that hominids had branched off from the lines leading to modern monkeys and apes about 5 million years ago. This was much more recently than most paleontologists believed, so it took a long time for fossil experts to accept this new molecular approach to studying evolution. In the meantime, paleontologists have recognized that the two methods are much more powerful when they are used together. ("The Search for the Earliest Hominid" earlier in the chapter discusses some of the reasons why fossils and DNA give different dates for evolutionary events.)

Applying the molecular clock method to humans was tricky because it worked best when there were easily measurable differences between the DNA of two samples. In the case of chimps and humans this was fairly straightforward, but the surprising genetic uniformity of modern *Homo sapiens* made it hard to apply to humans. And at the time, obtaining DNA sequences was still difficult and expensive. Later the problem would be solved with the discovery of parts of the human genome that evolved very quickly (described in chapter 3, "DNA Fingerprinting"), as well as new techniques that made it much easier to sequence DNA. But Wilson got around the problems with a brilliant idea. Instead of looking at the cell nucleus, he decided to focus on another type of DNA found in cells: the genes of mitochondria.

These cellular structures were introduced at the beginning of the chapter ("The Origins of the Genome") in the context of Lynn Margulis's hypothesis that they began as bacteria that invaded an ancient cell. One strong piece of evidence for this is the fact that mitochondria, which lie outside the cell nucleus, have their own collection of DNA and reproduce independently of the cell's genome. Most cells have more than 1,000 copies; because mitochondria have identical sequences, a lot of identical DNA can be obtained from a single cell. This is important when looking for genes in decayed samples, such as fossils. Wilson was also interested by the fact that some regions of mitochondria undergo mutations much more quickly than DNA in the nucleus—about 10 times as fast. And they have another curious feature: People inherit all of their mitochondria—along with their genes—from their mothers.

The reason lies in the way an egg and sperm fuse during fertilization. When a sperm penetrates the egg, its nucleus is combined with that of the egg, bringing their DNA together. This creates a new combination of chromosomes and a unique new genome. But only the sperm's nucleus is used—the rest of the cell is destroyed, including its mitochondria. The egg, on the other hand, comes equipped with a large collection of mitochondria; as the cells in which they live divide to form a body, the mitochondria take up residence in the body's cells.

Men have them too, of course—but they inherit them from their mother. This means that if a man's mitochondria undergo new mutations, those changes die with him, whereas changes in his wife's will be passed along to their children. Her female children will also pass them along, on and on, as long as her descendants continue to give birth to girls. A mutation creates a unique spelling for the gene, like the serial number on a marked $100 bill. If the bill is stolen from a bank, its number permits it to be traced back to its source.

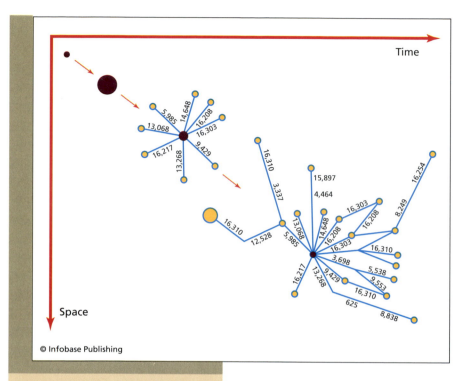

© Infobase Publishing

The X chromosome and human migrations. Researchers have tracked the migrations of human groups by studying mutations in 16,569 bases that make up mitochondrial DNA. The circles in this illustration are historical women with a particular sequence; the lines represent groups that have split off and undergone new mutations (which are shown by numbers that represent the position of the mutation in the sequence). Each line represents a group with a new mutation that has split off. The large cluster at the bottom right represents a population in Papua New Guinea.

This makes the inheritance of mitochondrial genes something like the female version of a Y chromosome, which is only passed from men to their sons. By studying this DNA in people all across the world, and using the concept of the molecular clock, Wilson's lab at Berkeley was able to trace each sample back to a common ancestor, a woman who likely lived about 140,000 years ago. Wilson's first estimate was actually somewhat older, placing her birthdate at about 200,000 years ago. A more recent analysis led to the revision.

What was so special about Eve? She was not the only female alive at that time. She was not the first member of the human species to evolve, or the only survivor of some worldwide catastrophe. In fact, there might not have been anything special about her at all, other than the fact that she probably had lots of children. The only thing that can be stated for certain is that she has an important place in the genetic history of *Homo sapiens.* Scientists have taken DNA samples from people living all over the world—from Australian aborigines to natives of the remote jungles of South America—and every one of them has specific "markers" from Eve's mitochondrial DNA.

To understand this, think once more about the mitochondria in men's cells. If a woman has only sons, she will pass her mitochondrial DNA to them, but that is the end of the line because they cannot pass the genes on to their own children, whether they be sons or daughters. So in a single generation any unique mutations in those genes will be gone. If she has daughters, her mitochondrial DNA will survive as long as they also have daughters.

Over the past 140,000 years, Eve's daughters have continued to bear females, whereas many other lines have died out, giving birth only to sons or leaving no offspring at all. People also have markers from other women throughout history, including other females alive at the time of Eve, but her DNA is the only set to have been inherited by everyone. She is the only common link in every human's family tree.

Theoretically, researchers realized, the same approach could be taken with the Y chromosome to find an ancient "Adam" who sired an uninterrupted line of sons. This approach worked fine when comparing DNA sequences from chimps and humans because there were plenty of markers to compare, but for many years researchers were unsure whether it could be done for the "nuclear" DNA within one species because there might not be enough variation in genes. It seemed that the Y chromosome, which had a lot of material that used to be genes but had become useless, ought to undergo mutations all the time. However, the first studies of regions of the Y chromosome revealed a maddening similarity, even though men were picked from all over the globe. Very few mutations were found; in fact, some regions of the chromosome were perfectly identical. As it turns out, the Y chromosome is the part of the genome that has changed the least over time.

But in 1985 the laboratory of Marcello Siniscalco, an Italian geneticist who established the Department of Human Genetics at the University of Leiden, in the Netherlands, found one region on the chromosome that did vary between individuals. Eventually its analysis allowed geneticists to find a male counterpart to Eve. "Y-chromosomal Adam" lived between 60,000 and 90,000 years ago. He never met Eve, but his descendants probably met a lot of hers.

These two individuals must have existed, and scientists are becoming more confident about the date provided by the molecular clock, but pinpointing where they lived is more tricky. To find out, scientists have had to cluster people into groups that represent major branches of Eve's and Adam's family trees. Bryan Sykes, a professor of genetics at the University of Oxford, Great Britain, has used mitochondrial DNA to identify seven clusters that make up the current population of Europe. Each of these subgroups, called *haplogroups,* can be traced back to a single individual—like the founder of a major clan—who was a descendant of Eve. All modern Europeans can trace their ancestry back to one of these women. In his 2002 book, *The Seven Daughters of Eve,* Sykes gives the founders of the haplogroups names and even fantasizes about what their lives may have been like.

Using molecular clock techniques, Sykes and other researchers can estimate when particular groups branched off from each other, and this is important in trying to discover where they originated. The idea goes something like this: Suppose every

(opposite page) The Y chromosome and human migrations. Mutations in regions of the Y chromosome can be used to reconstruct patterns of migration across the globe and to estimate the times at which particular splits occurred. (The same can be done with mitochondrial DNA.) This chart shows the origins and spread of each of the major human haplogroups. All males alive today had a common ancestor who lived in eastern Africa more than 60,000 years ago. *(Modeled after FamilyTree DNA, ©2006)*

human being in the world inhabits a small town on the coast of an ocean. After many centuries, the town becomes overpopulated and small groups begin moving inland, creating many new

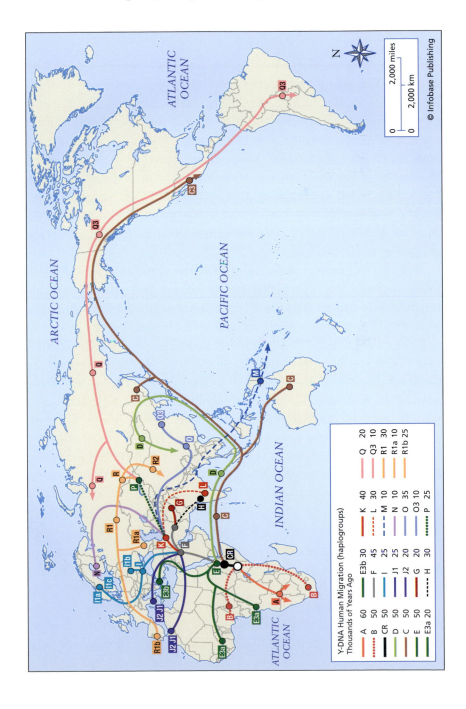

ATLANTIC OCEAN

ARCTIC OCEAN

PACIFIC OCEAN

INDIAN OCEAN

ATLANTIC OCEAN

N

0   2,000 miles

0   2,000 km

© Infobase Publishing

Y-DNA Human Migration (haplogroups)
Thousands of Years Ago

A	60	K	40	Q	20
B	50	L	30	Q3	10
CR	50	M	10	R1	30
D	50	N	10	R1a	10
C	50	O	35	R1b	25
E	50	J2	20	O3	10
E3a	20	J1	25		
E3b	30	I	25	P	25
F	45	G	20	H	30

towns. As each group leaves, it takes along the versions of the genes that currently exist in the town. When it arrives at its new home, the genes of those who left and those who remained slowly begin to diverge from each other. After a long period of time, their genes will be significantly different than those of the homeland.

Now suppose that wave after wave of people leave. The genes of the group that leaves the latest will be the most similar to those of the hometown. This gives scientists a molecular clock to detect different waves of emigration. The population of each new place acquires unique changes in its genome, a process known as *genetic drift*. Each place starts a new branch of the genome that evolves in a slightly different direction from the others. And eventually new settlements also outgrow their towns and spawn new waves of emigration. This results in a map of the world full of new cities whose inhabitants have unique genetic markers.

In this scenario, suppose that after tens of thousands of years, people have forgotten where they came from. At this point along comes a group of scientists who want to find the ancestral hometown. They can do so by checking the genes of people in each village. Since each of the younger villages was established by a subset of the original population, the scientists find only one branch—or a few branches—of the genetic tree among its residents. This is different from the original village, which was the only place that ever served as home to all of the branches. And a few of the old established families might have stayed in the city of origin and had children who never emigrated. These things mean that—barring some disaster—the genetic profile of the ancestral village will be different from any of the younger ones. The people who live there will share the oldest markers with the first wave of emigrants, but they will not have the changes that occurred since this group departed. They will bear traces of all the major branches that left the village at a later date. And they may have genetic markers that are not found anywhere else, because the genomes of families that never moved also underwent unique changes (that were never dispersed). Even if in modern times large numbers of peo-

ple have moved from place to place, the patterns will likely be detectable through a statistical analysis.

When this procedure is applied to data from mitochondrial DNA or the variable regions of the Y genome, it reveals that both Eve and Adam likely lived in Eastern Africa, in an area cur-

!Kung bushmen in a shelter. The !Kung inhabit a region that many researchers believe to be the place of origin of modern humans. Many !Kung still practice the hunter-gatherer lifestyle that was the basis of human society for most of history. *(Mike Scully)*

rently located within the country Ethiopia. Unlike populations anywhere else in the world, the genomes of African Bushmen from tribes called the !Kung and Khwe have DNA representing the oldest human lineage and other extremely old *haplotypes*. They also have more genetic variety than any other groups living in Africa. This is exactly the profile that scientists would expect to find in the group that spawned of modern humans. Genetic evidence suggests that a wave of emigrants left approximately 100,000 years ago to populate the world. They spread to become the ancestors of everyone alive today.

POPULATING THE GLOBE

From their birthplace in Africa, modern humans spread to Asia, Europe, and the rest of the world. The same genetic methods that have led to hypotheses about their origins are now permitting researchers to track the migrations of the children of mitochondrial Eve and Y-chromosomal Adam. Some of the routes taken by early *Homo sapiens* have become clear. The account here is one that has been assembled by Stephen Oppenheimer, a medical researcher at Green College in Oxford, in Great Britain. It is based heavily on findings by Allan Wilson and the Italian researcher Luigi Luca Cavalli-Sforza (1922–), long-time geneticist at Stanford University, as well as evidence from many other laboratories, the fossil record, and archeological finds from across the world. The story can be seen as an annotated map on the "Journey of Mankind" Web site from the Bradshaw Foundation (see "Further Readings"). Not all experts agree with the details, and there remain a number of mysteries to solve involving both fossils and genes. But these multidisciplinary studies are filling in many of the gaps.

One of the most interesting mysteries concerns the extinction of earlier species of hominids. The Neanderthals, for example, originated in Africa and spread widely through Europe and Asia. But after the arrival of modern humans they vanished without a trace, and so far no one has been able to explain why. Several explanations have been proposed (see "What Happened to the Neanderthals?"), but researchers have not yet reached a conclusion.

One mystery about the settlement of the globe involves the diversity of the human species. The first modern humans lived more than 150,000 years ago, according to most paleoanthropologists and geneticists. That is enough time to have generated

What Happened to the Neanderthals?

Neanderthals lived throughout the area that is modern Europe, Asia, and many other parts of the world from about 300,000 to 30,000 years ago. They used tools and carried out ritual burials, and during much of this time they lived alongside modern humans. Then they completely disappeared, and no one knows why. There are two main hypotheses: Either new groups arriving from Africa "absorbed" older hominid branches through interbreeding—which means that older forms somehow gradually became newer ones—or modern man replaced them.

The first hypothesis seems plausible because in modern times, humans who move to a region usually blend with the people living there. After wars or other types of conflicts, in the short term the groups may remain separate, but over hundreds or thousands of years they intermingle. The situation of modern humans and Neanderthals or other older hominids may be different because they belong

A replica of a complete Neanderthal skull. Until recent work analyzing DNA obtained from Neanderthal fossils, scientists were limited to comparing their anatomy to modern humans and other fossil hominids, leaving open many questions about their origins and fate. (*Dapper Cadaver*)

(continues)

(continued)

to different species. On the other hand, it might have been possible for them to mate and produce hybrid offspring. Chimps and bonobos—another type of chimpanzee—are separated by about 2 million years of evolution, but they can still mate. Lions and tigers can also mate and produce fertile offspring. However, horses and donkeys cannot.

If any of this had happened, scientists ought to find evidence of conflicts or cultural exchange between Neanderthals and modern humans. But while the two groups occupied the same regions at nearly the same time, archeologists have discovered almost no signs of contact. One exception is a site in central France, a cave called La Grotte aux Fées. Over the years French archeologists had found numerous Neanderthal remains in the region. But the cave also held artifacts from modern humans. In 2005 Paul Mellars, an expert in prehistory and human evolution from the University of Cambridge, Great Britain, carried out a careful dating of the layers of soil that had been excavated from the cave. He discovered that it was occupied by Neanderthals during a mild period in the last great ice age, from about 40,000 to 38,000 years ago. Then the weather grew colder, and modern humans likely moved down from the north, staying for at least 1,000 years. When things warmed up again, the modern humans moved on and Neanderthals once again occupied the cave. In an interview with ABC news, Mellars stated that, "This is the first categorical proof that Neanderthals and modern human beings did overlap in France for more than 1,000 years." Yet the archeological layers of the site were distinct; there was no evidence of intermingling between the groups.

The intermingling hypothesis makes two more predictions: that traces of Neanderthal genes should be found in modern humans, and that different "brands" of humans should have arisen in each region, reflecting a mixture of the new and old populations. Some fossils, such as remains from Lagar Velho in Portugal, seem to have characteristics of both Neanderthals and modern humans. Erik Trinkaus, a Neanderthal expert at Washington University in St. Louis, makes the same claim for modern human remains found in Romania. In an article in the May 1, 2007, edition of the journal *Proceedings of the National Academy of Sciences*, Trinkaus states, "Early modern Europeans reflect both their predominant African early modern human ancestry and a substantial degree of admixture between those early modern humans and the indigenous Neanderthals."

DNA extracted from fossils by Svante Pääbo, director of the department of genetics at the Max Planck Institute for Evolutionary Anthropology in Leipzig, Germany, reveals that Neanderthals were not direct ancestors of modern humans.

Some early claims of interbreeding had political or racial motives. Proposing that humans living in certain parts of the world were hybrids with "more primitive" hominids was used as an excuse to discriminate against them. It was also regarded as a potential explanation for differences between races. But studies of people across the globe have revealed that "race" is a cultural and historical phenomenon rather than something that can be described genetically. On the average there are more

(continues)

(continued)

similarities between people of "different races" living in the same region than people of the same race who live on different continents.

In 1997 Svante Pääbo's laboratory at the Max Planck Institute for Evolutionary Anthropology, in Leipzig, Germany, obtained samples of mitochondrial DNA from the first Neanderthal skull ever found. Markus Kring and other members of the lab found such major differences between this sequence and any modern human DNA that it closed the door on the possibility that Neanderthals had been the ancestors of modern humans; the scientists also believe it is unlikely that there was any interbreeding between the two groups. (If there was, their offspring may have been infertile, or their families have died out.)

Further studies of samples from other Neanderthals, carried out by Pääbo and other labs throughout the world, confirm the original findings. The researchers have also obtained nuclear DNA from the fossils. In 2006 they launched a project to sequence the entire Neanderthal genome in collaboration with the 454 Life Sciences Corporation of Branford, Connecticut. Some of the results have already been made known. Neanderthals possessed a version of the FOXP2 gene found in modern humans, which is thought to have played a key role in the evolution of language (see "The Evolution of the Human Brain"). When the genome is finished, the question of interbreeding will probably be answered for good.

If Neanderthals were not absorbed, they must have been replaced—but how? There is no evidence of large-scale war or other direct conflicts between the groups, and it is hard to understand why hominids such as Neanderthals completely vanished, everywhere, through "milder" forms of competition. Until modern humans they were probably the most intelligent species that had ever lived. But mathematical models show that if the two groups were directly competing for food or other vital resources, even a small difference in their rate of reproduction—from 1 to 2 percent—could have caused the extinction of the "slower" group within a very short time. Even so, the evidence provided by these methods is indirect. Most models of population genetics are based on studies of flows of genes within a single species. The theory of evolution explains how one species can split off into two, or how a new species arises from an existing one, based on slight differences that give some individuals a reproductive advantage over others. The American paleontologist and evolutionary theorist Stephen Jay Gould (1941–2002) believed that the theory could also account for competition between species, but experts on evolution remain divided on the issue.

Another possibility is that modern humans brought along diseases to which they had adapted, but against which the Neanderthals had no resistance. Viruses and bacteria can jump across species—most of the deadliest human plagues have originated in domesticated animals or creatures in the wild—and it is likely that many of the microorganisms that infected modern humans would also have infected Neanderthals. There are plenty of historical examples in which contacts between separate human cultures have led to the near destruction of one group: When the first group of Spaniards arrived in Central America they brought smallpox; historians estimate it unleashed an epidemic that killed a vast number of Native Americans. Smallpox also led to the deaths of nearly half the Australian aborigines when European settlers landed in 1788.

But all of these ideas remain hypotheses. The real reason for the demise of the Neanderthals remains a mystery—for now.

a lot of diversity in the genome, assuming some groups left Africa early enough, went their separate ways, and stopped interbreeding. But a look at the most diverse humans on the planet tells a different story. Even though the human population is huge (about 6.7 billion in July 2008) and inhabits every corner of the globe, any two people differ from each other only by about one in every thousand letters of the genetic code (0.1 percent). That is five times less than the amount of variation among chimpanzees, although current estimates place the world chimp population at just 125,000, and they live in much more restricted habitats. Why, then, is there less diversity among humans?

One hypothesis is that as recently as 75,000 years ago, humans went through a "bottleneck" that reduced a large population of modern humans to a small number, possibly as few as 10,000 people. This would have sharply diminished the overall diversity of the human genome because only the alleles of these 10,000 people would have survived. In addition, if the majority of those people were fairly closely related to each other, the effects on human evolution might have been extreme. Small populations are much more vulnerable to disease—the odds are lower that some people will have a chance mutation that helps protect them. They are also more likely to become extinct for other reasons.

What happened to everyone who did not survive the bottleneck? There was not necessarily a disaster; all this bottleneck really means is that everyone alive today is a descendant of one of those 10,000 people, and the rest did not pass along their genes. Another possible explanation is that a few thousand men and women simply had far more children than the rest, and their descendants were more fertile for hundreds or thousands of generations, outbreeding members of other family lines. But a worldwide disaster is possible, and in 1998 Stanley Ambrose, a professor of anthropology at the University of Illinois, discovered a plausible scenario by which it might have happened.

Lake Toba is a huge body of water located on the northern Indonesian island of Sumatra. It is the largest volcanic lake in the world, stretching 60 miles (100 km) in length and 30 miles (50 km) in width. It is also the site of a massive eruption that occurred between 67,000 and 75,000 years ago—scientists believe it is the most violent eruption of the last 25 million years and the second largest in 450 million years. The geological record indicates that it caused a "volcanic winter" in which summer temperatures may have dropped more than 21 degrees Fahrenheit (12 degrees Celsius) over a period of up to six years. The world was plunged into an "instant ice age" which lasted for 1,000 years. Ambrose believes that the Toba eruption had a devastating effect on life in the region and across the planet; it may have nearly caused the extinction of *Homo sapiens.*

At the time Southeast Asia was an important region in the spread of modern humans. Long before they had ventured from their original home in eastern Africa to the west and south, populating large regions of southern Africa. About 85,000 years ago a group began a journey eastward, crossing the mouth of the Red Sea and following the southern coastline of the Arabian Peninsula toward India. "All non-African people are descended from this group," Stephen Oppenheimer says—their children and grandchildren would go on to populate Asia, the Middle East, Europe, and finally the Americas. By 75,000 years ago descendants of this original exodus had settled Indonesia to the south and begun moving north along the Eastern coast of China.

Then came the Toba eruption. It likely led to the deaths of huge numbers of humans directly—through fallouts of ash and lava and massive tsunamis striking the coastlines of the entire region. Many regions of today's India and Pakistan were covered in up to five meters (over 16 feet) of ash. This and the abrupt climate change probably killed many more within a very short period of time through droughts and famines, in a massive loss of plants and animals that humans needed to survive. A few tropical regions offered a haven for some groups, but Ambrose estimates that a maximum of 15,000 humans survived, perhaps even fewer.

Over the next 10,000 years populations grew again and people began island-hopping, moving by boat to settle Australia and New Guinea. About 65,000 years ago temperatures began rising dramatically, opening up new areas in the north. Groups

Lake Toba, on the island of Sumatra in Indonesia, is the crater left over from a monstrous volcanic eruption that took place between 67,000 and 75,000 years ago and may have nearly caused the extinction of modern humans. (*NASA*)

moved into the Middle East and Europe. They had established a strong presence there by 45,000 B.C.E., beginning an expansion along the northern Africa coast, while other groups settled central and northeast Asia. Between 40,000 and 25,000 B.C.E. descendants of these populations moved into the arctic, including northern and eastern Russia.

North and South America had yet to be settled. A great land bridge connected today's Alaska and the Aleutian Islands to northeastern Asia, which scientists believe settlers began crossing between 25,000 and 22,000 B.C.E. One group seems to have traveled across what is now the northern United States to the eastern seaboard—a trip that might have taken several thousand years. Bones and other remains of this group have been found at a site called the Meadowcroft Rockshelter, located in southwestern Pennsylvania; they may be the earliest confirmed signs of the presence of humans in the New World. As their ancestors were moving eastward, another group moved down the western coast.

The period from 22,000 to about 15,000 B.C.E. saw the most extreme phases of the last great ice age, which chased most humans out of the northernmost regions of Europe, Asia, and North America. A few isolated groups remained in areas such as Alaska, Western Canada, and Siberia, but the rest migrated into warmer climates. The land bridge to the Americas became an impassable, frozen wasteland. The interior of the continent was populated by groups that had thrived on the eastern coast and now migrated inland.

Starting about 19,000 B.C.E. there were extensive migrations to the south, through Central America and along the eastern seaboard of South America, in what is currently Brazil. Regions that had been abandoned in the far north were inhabited again. New waves of nomads crossed the land bridge from Asia and traveled steadily southward, establishing settlements along the entire Pacific coastline of the Americas by about 12,500 B.C.E. By 8000 B.C.E. the Ice Age was over, the ocean rose to cut off the New World, and humans moved into nearly all of the areas that are settled today.

Until the late 20th century, piecing together human history—from its earliest evolutionary origins to modern times—was mostly the work of paleoanthropologists, archeologists, and climatologists. Through studies of the genetic code, molecular biologists and geneticists entered the picture. They learned to reconstruct deep evolutionary history by comparing the genomes of very diverse organisms, and recent human history by studying DNA sequences that mutate at unusually high rates. Another method to study events of the past 100,000 years or so is to investigate DNA obtained from fossils. These methods have given insights into a number of questions such as the evolutionary relationships between humans, Neanderthals, and chimpanzees. The same techniques that have permitted scientists to pinpoint eastern Africa as the cradle of *Homo sapiens* and to track human migrations across the globe are allowing them to address some much more modern mysteries, and that is the subject of the next chapter.

3

Using Genetics to Solve Ancient Mysteries and Modern Crimes

In 1988 a young British geneticist named Alec Jeffreys (1950–   ), working at the University of Leicester in Great Britain, was studying the evolution of a molecule involved in providing oxygen to muscle tissue. Jeffreys and his colleague Polly Weller were comparing versions of the myoglobin gene found in chimpanzees and humans. Like most human genes, myoglobin contains extra regions of code called *introns* whose information is not used to produce proteins. Jeffreys and Weller noticed something interesting: One of the introns contained a 33-letter sequence that was repeated four times in a row. They wondered whether the genome held any more copies of the sequence, so they went on a "fishing expedition" to find them. They discovered that scattered throughout the chromosomes are hundreds of small, repeated bits of code. There was something odd about them. While the repeats were found in the same locations (called *minisatellites*) in people's chromosomes, the number of repeats varied highly from individual to individual. Closer study revealed that these regions evolved so quickly that every person (except for identical twins) had a unique profile. Most were inherited from a person's parents, but a few would be unique—like a "DNA fingerprint."

Alec Jeffreys, a geneticist at the University of Leicester in Great Britain, invented "DNA fingerprinting" in the late 1980s. *(NIH)*

Jeffreys immediately realized that the rapid pace of minisatellite change could be used as a sort of "bar code" to match samples of DNA to a particular person. It could also be used to determine family relationships, as in matching a child to his or her parents. The first practical application of the method was to prove that a young boy who had immigrated to Great Britain from Ghana was really who his mother claimed he was. Since then DNA fingerprinting has become a standard tool in medical research and forensics. It has been used to solve thousands of crimes and find answers to some fascinating historical mysteries. This chapter presents a few of them.

ÖTZI'S GRANDDAUGHTER

It is one of the "coldest" cases in criminal history. The corpse of "Ötzi the Iceman," a Bronze Age hunter who died in the Alps, went undiscovered for more than 5,000 years. For a decade after his discovery in 1991, Ötzi was believed to have died of natural causes. But in 2001, when X-rays and a CT scan revealed an arrowhead lodged in the body's shoulder bone, investigators began to wonder whether he might have been murdered. Analysis of his DNA and other biological substances found with the body have provided scientists with some fascinating insights into his life—and also the names of some of modern relatives.

Ötzi was discovered by two German tourists as they hiked along a glacier in the Ötztal Alps on the border between Austria and Italy. They saw the decomposed body of a man protruding from the ice, lying facedown. When the local authorities came they removed the body and sent it to the morgue of the nearby city of Innsbruck—only then realizing that they were dealing with a mummy that was several thousand years old.

A number of artifacts were found alongside the body, and studying them has provided a wealth of information about life in prehistoric Europe. Ötzi was dressed in layers of clothing made of deer and goatskins and a bearskin hat. His shoes were made of goatskin folded around the top of the feet and sewn to a bearskin sole, with a bed of grass for warmth. He carried a backpack, medicinal plants, and a fire-making kit with flints to strike sparks and a piece of coal to get a fire started quickly. He also had weapons: a copper axe, an unstrung bow, 12 blank shafts for arrows, two broken arrowshafts, unfinished arrowheads, and a dagger. Each of the objects has been intensely studied by archeologists and materials scientists to discover how it was made and where it originated.

Several laboratories have been studying the DNA of Ötzi, the remains of food found in his stomach and intestines, and traces of blood on his possessions to discover what they can about his life and his final hours. The arrow lodged in his shoulder might have been the cause of death. An expert in prehistoric weapons, archeologist Thomas Loy of the University of Queensland, in Brisbane, Australia, found considerable additional evidence that the last days of Ötzi's life had been filled with violence. One of his own arrows bore traces of blood from two other people—making it likely that he had shot them and retrieved the arrow. The shaft was probably broken when a third shot missed its mark. His knife was stained with the blood of yet another person. More traces of blood, from a fourth individual, were found on the left side of his coat. Loy and his colleagues believed that he may have been carrying a friend.

A study of Ötzi's teeth and bones have provided clues about his origins. Geologist Wolfgang Mueller, at the Australian National University, in Canberra, studied the makeup of one of the body's leg bones and the enamel of his teeth. These parts of the body capture a signature of the location where a person lives. The minerals of each region create a unique profile of isotopes—charged versions of the elements—that become deposited in the mineralized parts of the body. Reporting in 2003 Mueller said, "From the enamel it is possible to reconstruct the composition of the water Ötzi drank and get clues about the earth where his food was grown. As a result we now know Ötzi came from near to where he was found from the Eisack Valley. He spent his childhood there. And he spent his adulthood in lower Vinschgau." Both of these areas lie in the Italian Alps, just south of where the body was found.

Several laboratories have participated in the analysis of Ötzi's mitochondrial DNA. An original analysis carried out in 1994 by the lab of Martin Richards, a geneticist at the Institute of Molecular Medicine of John Radcliffe Hospital, in Oxford, Great Britain, and Svante Pääbo's laboratory in Munich found similarities between the iceman's DNA and people currently living in the Alps. The groups retrieved a sequence 352 bases long and compared it to the same region of DNA in hundreds of people around the world. It most closely resembled the genetic code of 88 modern inhabitants of the Alps, differing from theirs only by an average of 3.38 letters. The sequence of one of these subjects, a person who lived nearby, proved to be an identical match. Inhabitants of more distant regions of Europe were nearly as similar; 255 Northern Europeans differed on the average by 3.73 letters. Even in this group there were nine people with identical sequences.

In 2005 the laboratory of Franco Rollo, an expert in ancient DNA at the University of Camerino, Italy, was able to obtain a better sample from the mummy's intestinal tissue and analyze a longer sequence of its mitochondrial DNA. The information included the sequence that had been analyzed 11 years earlier but also held part of the protein-encoding region of a gene. Since these regions undergo mutations at a lower rate than noncoding DNA sequences, Rollo and his colleagues hoped the study would be able to narrow down Ötzi's relationship to modern people and ancient European groups.

The analysis was successful. The scientists discovered that Ötzi belonged to one of the nine haplogroups that make up the current European population. These are populations that share the same haplotype, a common set of genetic markers that are inherited together. (Haplogroups based on mitochondria and sequences on the Y chromosome were introduced in the previous chapter; see "Mitochondrial Eve and Y-Chromosomal Adam"). Ötzi was a member of group "K," which Oxford researcher Bryan Sykes has imaginatively called "the daughters of Katrine." Katrine is the fictional name of a real person—a "clan mother" who lived about 12,000 years ago and is the ancestor of every person in the haplogroup. Rollo's work made the identification even more precise by showing that the K group split into two main branches. Ötzi, like many other Europeans, belongs to K1, but his DNA has some unique spellings. Sykes, who has been working on the haplogroups for many years, originally believed he had found one woman living in Great Britain whose DNA sequence is identical to Ötzi's. However, additional sequences have revealed that she is not his direct descendant; in fact, no one alive today may be.

THE ORIGINS OF THE BASQUES

Between France and Spain lies a rugged, toothlike chain of mountains called the Pyrenees. Here, in a border region between southwestern France and northern Spain, is the homeland of the Basques, a group whose origins and culture have long been a puzzle to historians. For most of modern history, the people of this region have maintained their independence in spite of attempts to integrate them into the Roman empire, France, Spain, or various kingdoms that have continually arisen and redrawn European boundaries. The Basques' special character owes a great deal to the fact that the Romans had little interest in the region. Their language and culture were not overwhelmed by Latin, and the people were granted a degree of independence because of their usefulness as soldiers in wars against Rome's enemies. When the Roman Empire collapsed, the group found itself between new fronts: Celts, Franks, Moors, and Castilians. There were new groups to whom they could ally themselves in search of true independence.

One reason that they stand out is the Basque language, the uniqueness of which caught the interest of 19th-century experts in the new science of philology. This field was originally devoted to the study of ancient texts, but the focus was increasingly shifting toward the history of languages. Early philologists included Jacob and Wilhelm Grimm, authors of *Grimm's Fairy Tales.*

The Grimms and others realized that modern languages had obviously arisen from earlier forms, and that a single ancient language had sometimes evolved into several modern ones. Philologists discovered that Greek, Latin, German, Persian, Sanskrit, and the Slavic and Celtic languages had arisen from a much older common language that they called "Indo-European." Many dreamed of integrating Indo-European into an even larger tree that would include all of the languages of the world, descended from the extinct original language of mankind. But some modern tongues were simply too strange to work into the tree—including Basque. It was not an Indo-European language.

Where had it come from, and why did it stand alone, island-like, in the middle of Europe? Some linguists claimed it was related to ancient Iberian, once spoken throughout the Spanish and Portuguese peninsula but now extinct; others saw connections to languages spoken in Caucasia, a mountainous region which links Europe to Asia. Interestingly, this area played an important role in human history: Modern humans moved through it as they spread from Africa to settle Asia and Europe. Some hypothesized that the Basques might represent an ancient tribe that had somehow remained isolated and survived as other groups settled the rest of Europe. A few authors even suggested they were part of the original wave of *Homo sapiens* who emerged from Africa and settled Europe.

In the early 1960s geneticist Luigi Luca Cavalli-Sforza of Stanford University began combining genetics and linguistic methods to trace historical migrations of populations across the globe. "The challenging task of reconstructing the history of human evolution can hardly be entirely satisfactory using only evidence provided by the genetic data," he wrote in the 1994 book, *The History and Geography of Human Genes.* "Information from historical, linguistic, anthropological, and archeological sources is also useful, and it should be compared with the genetic evidence if we wish to reach fully satisfactory conclusions." But each of these subdisciplines provides a partial picture that is easily misleading. "Only genes almost always have the degree of permanence necessary for discussing fissions, fusions, and migrations of populations that took place during the history of our subspecies."

Cavalli-Sforza began working on the topic before the invention of DNA fingerprinting and before it was feasible to study the DNA sequences of mitochondria or the Y chromosome. But as Linus Pauling had shown, even variations in single molecules such as the protein hemoglobin could be used to study evolution (see chapter 2, "Molecular Clocks"). The discovery of the blood groups A, B, and O and the *Rhesus (Rh) factor* (which represent differences in the proteins on the surface of a person's red blood cells) had given researchers a starting place, but they did not provide high enough "resolution"; more genes would be necessary to produce a detailed map of the migrations of European populations. Even so, the Basque profile stood out within Europe. The blood types B and AB are practically nonexistent within the population, and proportionately, they have the highest rate of Rh-negative blood in the world. These facts suggested that the Basques had been isolated for a great deal of their history.

In 1963 Cavalli-Sforza demonstrated that "even with as few as 20 alleles from five genes one could successfully attempt a reconstruction of human evolution. . . . Later experience proved that a larger number is desirable or even necessary." One question that interested him was whether scientists could define races or ethnic groups through genetic markers. Either race was something "real"—in genetic terms—or it was a social and cultural phenomenon: People gained a fuzzy impression of one another based on language, culture, and physical characteristics, but in reality these were superficial differences.

Every genetic study had supported the second conclusion. Many systems of breaking down human beings into subtypes had been developed, but each yielded a different number of races—from three to 60, depending on the biases of the scientist. Nearly always, a wide range of alleles for each gene could be found in each population. Arbitrarily one might say that a race was a particular cluster of haplogroups, but the people in these groups looked different, spoke differently, and came from various cultures—in other words, the man on the street would say they belonged to different races. Genetically, race does not really exist.

As Cavalli-Sforza and his colleagues expanded the number of genetic markers that could be compared, including Y chromosome and mitochondrial DNA sequences, they found that four groups of Europeans stood out: the Basques, the Lapps of Northern Finland, the Icelandic people, and the inhabitants of the island Sardinia. The Basque language, Cavalli-Sforza reports, "is considered an isolate language, very probably a relic of a pre-Indo-European one. All possible relatives of Basques are very distant from it linguistically and geographically. Genetically Basques are European but are sufficiently distinct that they can be recognized as an isolate, probably proto-European."

Not that they have been completely isolated—there has been an influx of genes through marriage between Basques and members of other groups. The study showed, however, that this mixing happened at an unusually slow rate. In addition to retaining a language descended from an ancient tongue, one that predates the arrival of Indo-Europeans in Europe, they have retained some of their original genetic characteristics. With the arrival of the last ice age, pockets of modern humans

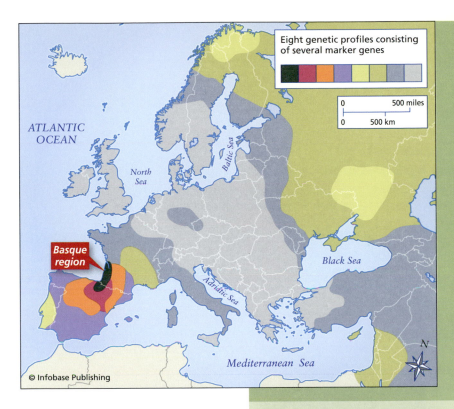

Eight genetic profiles consisting of several marker genes

ATLANTIC
OCEAN

North
Sea

Baltic Sea

Basque
region

Black Sea

Adriatic Sea

Mediterranean Sea

© Infobase Publishing

in southwestern France and the Iberian peninsula (Spain and Portugal) were likely cut off from one another for long periods of time. That led to the development of unique genetic and cultural characteristics. When the glaciers retreated, there was a huge influx of new immigrants. In most places this led to a thorough intermingling of the new and old populations and cultures. In the Basque

Europe and the Basques. Colored areas showing the locations of haplogroups mark this map of Europe. Luca Cavalli-Sforza has used this information to show that the Basques (who live on the border of western France and Spain) were probably one of the earliest groups of modern humans to arrive in Europe and that they have remained relatively isolated from other populations. The Basque language may be a direct descendant from an original European tongue, and the people have several original genetic characteristics.

country that process has happened much more slowly, which means that more traces of the oldest modern human to reach Europe can still be found there.

Making a Pedigree

A pedigree is a chart used by geneticists to reveal family relationships and patterns of inheritance. The goal is to track a gene, DNA sequence, a haplotype, or another inheritable feature through the generations of a family. This has several uses; it can show, for example, the chances that the children of a particular couple might inherit a genetic disease. A pedigree is a type of family tree that usually leaves out names—for reasons of medical privacy—but indicates the sex of each person, because of the importance of gender in heredity.

A standard set of symbols has been developed to draw pedigrees:

- Square = male
- Circle = female
- Diamond = child whose sex is unknown
- Horizontal line = connects people who had children together
- Vertical line = connects children to their biological parents
- Shaded = person who exhibits a dominant trait
- Empty = person who does not exhibit a dominant trait
- Dot = person who carries a recessive allele
- Strike-through = person who has died. This is important because death due to a disease might reveal something about a disease gene. Deaths due to accidents might occur before a person exhibits a disease or the genetic feature that is being studied.

Below is an example of a pedigree that accompanies a story later in the chapter. It represents the family of the

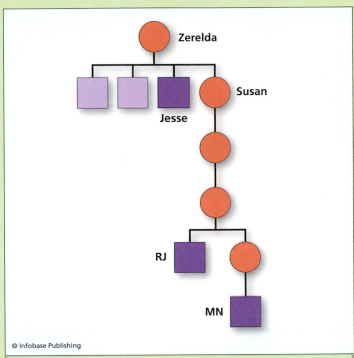

© Infobase Publishing

A pedigree of the maternal relatives of Jesse James starting with his mother, Zerelda. (Squares represent males and circles represent females.) A great-grandson (RJ) and a great-great-grandson (MN) of his sister Susan inherited mitochondrial DNA passed down from Zerelda, confirming that the body buried in the Mount Olivet, Missouri, cemetery is also one of her descendants.

19th-century outlaw Jesse James, and was used to establish whose body was really buried in his grave. The James pedigree begins with Jesse's mother, Zerelda James (his father is not depicted because he is not relevant to the information that the authors used). She had four children: three boys and a girl. Jesse's square is black because the

(continues)

study aims to trace one of his features: a sequence in his mitochondrial DNA. Both Jesse and his brother Frank had children, but they are not important to the study, because only women pass mitochondrial DNA to their children. The chart shows one line of his sister Susan's descendants to the present day. Susan had one daughter, who also had a daughter, who had one son and a daughter. RJ, the son, is still alive and contributed DNA to the study. His sister also had a son, MN, who likewise contributed DNA. Since both of these men stem from a maternal line stretching back to Jesse James's mother, they should have the same type of mitochondria as Jesse, and the point of the study was to establish whether this was true. Like most pedigrees used in studies, the chart leaves out any information that is not relevant. (The conclusions drawn from this chart come later in the chapter.)

Family relationships are only one type of question that can be answered through pedigrees. Knowing the makeup of a tree allows researchers to predict what features (such as a mitochondrial DNA sequence, or another genetic characteristic) a person ought to have. They can also be built backward: By assembling a wealth of data on a group of living people (such as Europeans, or all human beings), scientists can reconstruct the tree itself (as in the case of the Basques' place in European populations, or the branches starting with Mitochondrial Eve). Yet another use is to track down the genes responsible for disease, discussed in chapter 4.

THE HEIRS OF THOMAS JEFFERSON

Even if he had never served as the third president of the United States, Thomas Jefferson (1743–1826) would have gone down in history as one of the country's most brilliant figures: as the author of the Declaration of Independence, the architect who designed Monticello, secretary of state, governor of Virginia, and the young nation's minister to France. His personal life was marked by the tragedy of the loss of his wife, Martha, who died at the age of 34 after bearing him six children (all but two of whom died as infants). Jefferson never remarried, and in 1802, during his presidency, a scandal arose when a journalist accused him of having fathered a child with one of his slaves.

Sally Hemings (1772–1836) was a slave who had been owned by Martha's father, John Wayles; Sally may also have been his daughter, making her Martha's half-sister. When Wayles died, her family became the property of the Jeffersons. After Martha's death she stayed with Jefferson and accompanied the young statesman to France. While abroad she was free—slavery was illegal in France—and did not want to return to the United States. But she eventually consented to return when Jefferson promised to free her children when they reached adulthood. Many years later one of her sons told a newspaper that she had conceived a child with Jefferson in France. That infant died, but Hemings went on to have seven more children, all of whom, according to her son, were fathered by Jefferson.

One of the surviving children from Jefferson's marriage to Martha was a daughter, also named Martha, and her descendants have denied the story that he fathered any children with Sally. Instead, they say the real father was one of Thomas Jefferson's nephews, either Samuel or Peter Carr.

While many historians support the claims of the Hemings descendants, there has been some doubt about how many children the pair really had. The 1802 newspaper article concerned a possible first son named Tom, who was given up for adoption to a family named Woodson. There really was a Thomas Woodson, who went on to have 11 children. Two centuries later Woodson's descendants number about 1,400 people, many of whom believe the family legend that Thomas Jefferson was Tom Woodson's father.

Eugene Foster, a pathologist from Charlottesville in Jefferson's home state of Virginia, became interested in the story after his retirement. He had been thinking about the possibility of using DNA to try to resolve some of the questions, but studies of the family's nuclear DNA probably would not give a satisfying answer. This is a general problem with DNA fingerprinting. It does fine in establishing direct relationships between children and their parents, or matching a sample to the person it came from, but with more distant relationships things become unclear because so many ancestors have contributed to the genome. A colleague told Foster about the use of Y chromosomes to establish relationships between males. In 1996 he began looking for Jefferson or Hemings descendants and other potential relatives that would have inherited the same Y chromosome. He enlisted the help of Chris Tyler-Smith, a biochemist and expert in DNA at the University of Oxford in Great Britain.

The easiest test would have compared a Y chromosome directly passed down from Thomas Jefferson through Martha and Sally Hemings's descendants. But there were no male heirs on Martha's side. The alternative was to climb higher up the family tree, to Jefferson's father and grandfather. Their sons would have inherited the same Y chromosome as Jefferson and passed it down through their sons. Foster found a paternal uncle, Field Jefferson, who had a direct line of male descendants to the present day. Five of them provided DNA samples for the study. DNA was also available from males of the Woodson families and from John Weeks Jefferson, a descendant of Sally Hemings's son Eston. A final group consisted of descendants of Jefferson's nephews, the Carr brothers. Since they were his sister's children, they had inherited their Y chromosome from someone unrelated to Thomas. If one of the Carrs were the father, the differences would be obvious.

Foster and Tyler-Smith looked at a minisatellite region of the Y chromosome, the only area that undergoes enough mutations to permit distinguishing between families. They found a pattern of markers that were common to the men in Thomas Jefferson's lineage; it was also unique—it had never been found in other studies of Y chromosomes. The pattern was identical to that found in the Y chromosome of John Weeks Jefferson, the great-grandson of Eston Hemings, Sally's son. This meant, Foster concluded, that Jefferson was at least 100 times as likely to have been Eston's father than someone outside Jefferson's family. "This is a very conservative estimate and the true ratio may be much higher," Foster says.

On the other hand, there was virtually no chance that one of the Carrs had been the father. And as for the Woodsons' claims, there were such significant differences between their Y chromosomes and the Jeffersons' that Thomas Jefferson could not have been the father of Tom Woodson. This is supported by Jefferson's will, in which he freed only Sally, her children, and two of her other relatives.

THE GRAVE OF JESSE JAMES

Who is buried in Jesse James's grave? The question may sound like the beginning of a joke, but since 1882, when the legendary outlaw was shot and killed by Robert Ford in Saint Joseph, Missouri, there have been rumors that he survived. In 1947 an Oklahoma man named J. Frank Dalton came forward claiming to be the real Jesse James, now over 100 years old. He had engineered the death of an imposter, he claimed, to escape from the law. While almost no one took Dalton's claims seriously (he had also claimed to be a member of the Dalton gang and several other notorious celebrities), the circumstances surrounding James's death were confusing. To put the questions to rest, in 1995 a court ordered the exhumation of the body of the man shot in 1882, which had been claimed by the James family and buried on their farm in Kearney, Missouri. The body had been moved again in 1902, to the family's plot in Mount Olivet cemetery, also in Kearney.

James, his brother Frank, and their band gained notoriety after the close of the Civil War for a 15-year string of daring, bloody bank and train robberies throughout the midwest. In spite of the brutality of their crimes, in which a number of people were murdered, the Jameses achieved a sort of folk hero status during their lifetimes. There were several reasons, starting with the fact that the brothers had fought for the South during the Civil War, then as part of pro-slavery terrorist groups known as "bushwhackers" that ravaged Kansas and Missouri. After the war there was a great deal of resentment against the federal government. Robbing banks, stagecoaches, and trains was seen by some as a type of government resistance. Some of the bushwhackers turned to crime and began robbing banks across Missouri. Jesse and Frank James were part of these gangs, but their names first came to public attention with a robbery in 1869. The bank of Gallatin, a small town in northwest Missouri, was run by Samuel Cox, who had killed a notorious bushwhacker named Bill Anderson. Although Anderson had been an incredibly brutal guerrilla leader who had murdered hundreds of Union soldiers, including a group of unarmed soldiers on leave, he was regarded as a Southern hero. Newspapers played up the robbery as an act of revenge, a political statement—even though the gang failed to kill Cox and made off with worthless bank papers.

But the bold, daylight act called Jesse and his gang to the attention of John Newman Edwards, a fervent southerner who became editor of the *Kansas City Times* in the late 1860s. Newman did not need much convincing to see James as a patriot, a modern-day Robin Hood bent on avenging crimes against the South. He wrote articles that portrayed the criminal as a folk hero. (James had even begun leaving "press releases" at the scenes of his crimes, which Newman printed.) Newman even compared the James gang to King Arthur's knights of the Round Table. As years passed and the reconstruction of the South began to improve the situation, Edwards sank into alcoholism and neglected to publish later letters from the outlaw.

By 1882 the climate in Missouri had changed; southern heroes were no longer welcome. Businesses were hesitant to establish themselves in the "Robber State." Both Jesse and Frank married and went into hiding in Tennessee for a while; Frank was tiring of the outlaw life. But Jesse returned to Missouri to plan more robberies. He moved to Saint Joseph under the name of Thomas Howard.

During their absence a new governor, Thomas Crittenden, had been elected on a platform of promising to rid Missouri of the gangs. He collected money from the state's businesses to establish a huge reward, $10,000 for each brother. A member of Jesse's new gang, Robert Ford, cut a deal with the governor—he would kill Jesse and then receive the reward money and a pardon. Early one morning as he prepared to leave the house, James stopped to straighten a picture on the wall, and Ford shot him in the back of the head.

The body of "Thomas Howard" was claimed by the James family and buried in the front yard of the family farm, but the legends did not die easily; many people wished that he had survived. In 1995 James Starrs, a professor of law at George Washington University in Washington, D.C., obtained a court order to exhume the body. Starrs is a controversial figure who has been in the media spotlight for exhuming a number of other high-profile historical figures. In 2004 he participated in an exhumation of one of the victims of the "Boston Strangler," a person who killed 13 women in the Boston area between 1962 and 1964. The investigation revealed DNA evidence that matched neither the victim nor Albert DeSalvo, who had confessed to the crime. In 2007 Starrs participated in an effort to exhume Harry Houdini to clarify the causes of the magician's death (a motion that is still going through the courts at the time of writing). He has also hoped to examine the body of Meriwether Lewis—the famous explorer who, with William Clark, charted the American West. Starrs hoped to discover whether Lewis committed suicide or was murdered. Lewis may also have died as the result of an accident.

Jesse James's body was removed from the town cemetery in Kearney, Missouri, and Anne Stone and Mark Stoneking, researchers at the Department of Anthropology of the University of Pennsylvania, attempted to retrieve DNA from the bones. They were unsuccessful at first, probably because the soil was wet and slightly acidic, conditions unfavorable to the preservation of DNA. But one of the teeth eventually yielded usable DNA. And there was an additional sample: When the body had been moved from its original location at the James farm, some hair had been left behind. It was recovered in an excavation of the grave site in 1978 and had been preserved ever since in a Tupperware container. The mitochondrial DNA from the teeth and hair matched.

When the genetic code of the sample was compared to a large database of mitochondrial DNA sequences obtained in other studies from people in the United States and throughout the world, Stone and Stoneking found it had a unique spelling. They now needed to match it to a relative in James's maternal line. Starrs had tracked down a great-grandson and a great-great-grandson of Jesse's sister Susan (identified in the study as "RJ" and "MN"), who provided DNA samples for the study. Their mitochondrial sequences were a perfect match.

The study concluded that this meant one of three things: "(1) the exhumed remains are indeed those of Jesse James; (2) the exhumed remains are not Jesse James, but from another maternal relative of RJ and MN; or (3) the exhumed remains are from an unrelated individual who, by chance, happens to have the same mtDNA sequence as RJ and MN."

The last possibility is extremely unlikely, and the simplest explanation, of course, is that the body is that of Jesse James. But legends die hard. Conspiracy theorists on the Internet make a great deal of the fact that when an automobile salesman obtained a court order to exhume the body of J. Frank Dalton, hoping to test his mitochondrial DNA as well, the wrong body was dug up. So far, no one has found it necessary to try again.

THE FATE OF ANASTASIA, GRAND DUCHESS OF RUSSIA

One of the most famous missing-person cases in history—and probably the most famous case to be solved by DNA analysis—is that of Anastasia Nikolaevna Romanova (1901–18), one of five children of Czar Nicholas Romanov II and Czarina Alexandra of Russia. The story has inspired legends, novels, musicals, and films—even an animated movie. The czar and his family were victims of Russia's 1917 Bolshevik revolution, which installed a Leninist government to rule the country when the czar abdicated the throne. Initially, the royal family was placed under house arrest. But they still had the loyalty of many Russians who wished to restore them to power. In the midst of a civil war, with its hold on power still shaky, the new regime considered them a serious threat. During the night of July 17, 1918, the Bolshevik secret police murdered the entire family. But from the very beginning, rumors circulated that at least one of the daughters might have survived. Within a few years several women claimed to be Anastasia. It has taken 90 years for DNA evidence to finally close the case.

One reason Anastasia's fate remained a mystery so long was a missing eyewitness account of the execution and burial of the bodies. That information was included in a report filed by the police squad commander, Yakov Yurovsky, but it was hidden away in the state archives and only resurfaced in 1978. According to the report, the family had been quartered in a house in the city of Yekaterinburg, about 900 miles (1,500 km) east of Moscow. As an army of loyalists advanced on the city, guards woke the family in the middle of the night and told them to get dressed, on the pretext that they were being moved to a safer location. Instead they were herded into a room in the basement with a few servants, where they were all shot. Yurovsky wrote that a few of the family members survived the initial volley— bullets were deflected by precious jewels sewn into their clothing—but his soldiers finished the job with their bayonets.

The secrecy under which the executions were carried out encouraged rumors about a survivor. Some stories said that the Bolshevik police were raiding houses and trains, looking for Anastasia; that witnesses saw her alive with her mother and sisters in another city after the date of the execution; that two months after the murder, an injured girl claiming to be Anastasia was captured at a railway station. In most versions her survival was attributed to the help of a sympathetic guard. Over the next decades a number of women in Russia and elsewhere claimed to be Anastasia or her sister Maria. Some were participants in obvious frauds, aimed at milking relatives of the royal family for money, or making claims on the Romanov's estate (which had been seized by the Bolshevik government). But in the early 1920s, a young woman appeared in a mental hospital in Berlin, and for several decades she managed to convince a lot of people that she was Anastasia.

Anna Anderson (1896?–1984) first came to public attention after an attempt to commit suicide by jumping off a bridge in Berlin in 1920. She was rescued by a police official and taken to a mental hospital. Admission reports stated that her body bore the marks of bullet wounds and the scar of a Russian bayonet that had been plunged into her foot. She seemed to have lost her memory and spoke very little; when she did, it was with an Eastern European accent. A year later she told one of the nurses that she was Anastasia. She claimed to have been rescued by a Russian soldier named Alexander Tschiakovsky, with whom she had had a child; later records revealed that no one with this name had been part of the execution squad.

After being released from the institution, Anna Anderson began a long series of stays with Russian émigrés in Europe and the United States. She was visited by Russian and European royalty who had known the Romanov family. Most of them claimed that she could not be Anastasia. Yet others, including the son and daughter of the Romanov family doctor, who had died with them, made a positive identification. This was enough to keep the story rolling for decades.

If Anna Anderson was not Anastasia, who was she? A possible answer was provided by Doris Wingender, a young German woman who claimed that Anderson had lived with her family for several months. Wingender said that Anderson's real name was Franziska Schanzkowska and she came from Pomerania, a part of Prussia now located in Poland. During World War I she had worked in a weapons factory and had accidentally set off a grenade, badly injuring herself and killing a coworker. Afterward she suffered from mental illness and wandered in and out of various institutions. Schanzkowska disappeared in 1920, almost simultaneously with the appearance of Anna Anderson.

Anderson stuck by her story until her death in 1984, when her body was cremated. At about the same time new mitochondrial DNA and DNA fingerprinting were emerging as methods to establish identity. Researchers used the techniques to do two things: inspect the DNA of the Romanovs' remains, which had been found thanks to the rediscovery of Yurovsky's report, and compare it to Anderson's DNA. Samples from Anderson could also be compared with DNA from relatives of the Schanzkowska family.

The grave of most of the family members was excavated in 1991. It contained nine skeletons that turned out to be remains of the czar and his wife, three of their daughters, and four members of the royal family's household staff. The son and one of the royal daughters remained missing. DNA played a key role in the identification. Pavel Ivanov, a Russian forensics expert, took samples of DNA to Peter Gill, head of the British Forensic Science Service, in hopes that state-of-the-art British equipment would be able to extract useful DNA. Gill's lab obtained nuclear DNA from the bones and used DNA fingerprinting to prove that the three girls were the children of two of the adults in the grave. This did not prove, however, that they were the Romanovs—nor did it say which girl was missing—because none of the family's tissues were available for direct testing.

That was accomplished by comparing mitochondrial DNA from the bones with samples from relatives of the czar and czarina. Gill and Ivanov needed to find relatives who had inherited the same set of mitochondrial DNA—that is, someone related to the mothers of Nicholas and Alexandra, their daughters, or their granddaughters. For the czarina that turned out to be Prince Philip of Great Britain, the husband of Queen Elizabeth II. His great-grandmother, along his maternal line, was the mother of Czarina Alexandra. This meant that both had inherited mitochondria from the same source. If the DNA from the bodies of the woman and the three girls from the grave matched that of Prince Philip, it would mean they were maternally related. Gill's study reported that the sequences matched exactly.

Finding a match for the czar's DNA was more challenging. Two living relatives were found, and the sequences were identical to some of the mitochondria in the sample. But his cells held a second type of mitochondria with a different genetic code. This situation is rare but not unknown; it happens when the cells of a person's mother already have two kinds of mitochondria. The phenomenon usually disappears after just a few generations because one of the versions makes more copies and eventually dominates as new egg cells are made. That had already happened with the czar's relatives; they had no longer inherited both types of mitochondria.

The only alternative was to look elsewhere, so Ivanov obtained an order to exhume the body of the czar's younger brother Georgij, who had died of tuberculosis in 1899 and was buried in a cathedral in St. Petersburg. This time he took samples from the bones to the U.S. Armed Forces DNA Identification Laboratory, in Rockville, Maryland. The main purpose of this facility is to try to identify the remains of U.S. service personnel who have died during military conflicts. The mitochondrial DNA of the two brothers matched perfectly; Georgij, too, had inherited two types of mitochondria from his mother.

The tests proved that the grave held the remains of the czar, his wife, and three daughters—but two children were still missing. Could one of them truly have escaped—and might it have even been Anna Anderson? If a sample of her DNA could be found, it could be compared to that of the family. The cremation ruled out obtaining any DNA from her corpse, but five years before her death she had undergone an operation and the hospital had saved tissue samples. Another sample—hair that had been collected by her husband—offered a second source of DNA. In 1994 the material was sent to Gill, the U.S. army laboratory, and geneticist Mark Stoneking of Pennsylvania State University. DNA fingerprinting demonstrated that the hair and tissue came from the same person—Anna Anderson—and that she could not have been the daughter of the Russian royal family.

There were so many differences between her mitochondrial DNA and that of Prince Philip that the two could not be maternally related—that meant she could not be related to the czarina's mother, either. But a comparison to DNA from a man named Carl Maucher, a maternal relative of Franziska Schanzkowska, was a positive match. This makes it very likely that Anderson was a relative of the missing Polish girl—and the simplest explanation is that she was Franziska herself.

The Romanov pedigree, shown below, is more complex than that of the James family because it aims to answer two different questions: the identity of the females found in the grave

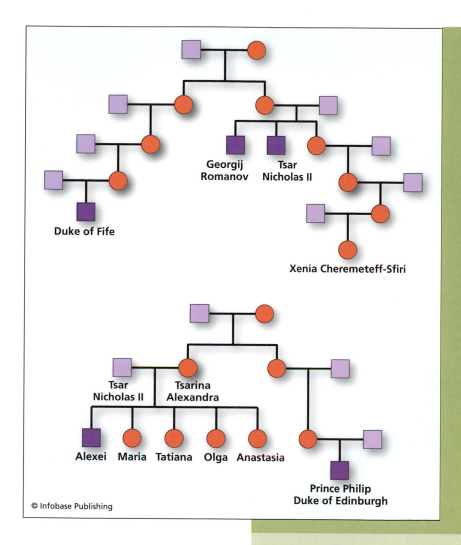

Georgij Romanov   Tsar Nicholas II

Duke of Fife

Xenia Cheremeteff-Sfiri

Tsar Nicholas II   Tsarina Alexandra

Alexei   Maria   Tatiana   Olga   Anastasia

Prince Philip Duke of Edinburgh

© Infobase Publishing

in Russia, and the identity of a male. Both depend on mitochondrial DNA. But since the man (presumably the father of some of the victims) and the woman (thought to be the mother) come from different families, different pedigrees are necessary to identify them.

Romanov family pedigree. Identifying the remains found in graves in Russia required matching mitochondrial DNA along two different matrilineal lineages. The top chart shows a line from Czar Nicholas's grandmother to her descendants, including the czar, his brother, and living relatives. The bottom shows a lineage from Czarina Alexandra's mother to her grandchildren, including Anastasia, and to Prince Philip, a living relative.

4

The Genetics of Health and Disease

Genetics was born with the discovery of laws that govern the passage of hereditary traits from parents to their offspring, but in the latter half of the 20th century the science took new directions. First, it began to reveal how instructions in the genetic code build organisms. Secondly, it gave researchers a powerful toolbox to alter those instructions and thus manipulate processes in cells and organisms. All of these aspects of genetic science are now helping scientists understand and treat diseases, which is a major focus of human genetics, and the main theme of this chapter. The main connection between genes and disease are the following:

- Infectious diseases are caused by parasites that take advantage of molecules and processes in the host as they invade cells. How ill a person becomes depends on whether a parasite can gain a foothold and how the immune system and other bodily systems respond. This explains why people with different versions of genes often respond differently to an infectious disease.
- Developmental defects occur when an embryo's genetic material causes it to grow in an abnormal way, suffer, or die early. These conditions may arise through new mutations as sperm and egg cells form, or they may be inherited from a parent. It is important to note that the difference

between normal and defective development is relative; what is unhealthy in one situation might be helpful in another. What begins as a detrimental mutation can become the basis of the evolution of new species.

- Cancer and some other conditions such as autoimmune diseases may begin with somatic mutations that occur in specific cells during a person's lifetime. The changes disrupt cellular processes like the timing of division, specialization, and migration. Often it takes more than one mutation to trigger the development of tumors. Some people have mutations that give them a greater chance of developing cancer; they already have a defect in a key gene, and it takes fewer mutations to launch the process of cancer.
- "Systemic" diseases, such as hypertension, cardiac disease, Alzheimer's disease, other degenerative conditions, and aging, arise from "inherent" genetic flaws. They usually appear in old age, after organisms have finished reproducing, so animals have not evolved defenses against these problems and are usually ill-equipped to handle them.

These categories are often closely linked to each other and to healthy processes. Sickle-cell anemia, for example, is a developmental disease that causes the body to build improper blood cells; it also protects people from malaria, an infectious disease. The Epstein-Barr virus, the papilloma virus, and other infectious agents sometimes lead to cancer. Tumors have also been linked to degenerative conditions and the process of aging, as the body loses its ability to recognize aberrant cells or repair errors in the genetic code.

Genetics promises to revolutionize medicine in two ways: through methods that can be used to identify the causes of diseases, and through the development of tools and therapies to try to improve the health of affected people. The main focus here will be the search for causes, but the chapter also briefly introduces some approaches to cures.

FINDING DISEASE GENES

Hundreds of diseases are now known to be caused by defects in genes, often a single misspelling in the genetic code. Many more health problems are thought to be the work of multiple genes. And some conditions involve regions of the genome that do not contain genes. There are two main ways by which researchers discover such connections. *Forward genetics* means observing that a disease follows a pattern of inheritance similar to a dominant or recessive gene and then tracking down the part of the genome that is responsible. *Reverse genetics* means interfering with a specific molecule in a mouse or another laboratory organism and then noting if it develops problems. If the symptoms resemble a human disease, the scientist can investigate tissues taken from patients to discover whether they have a defect in the human form of the same molecule. The development of new technologies since the late 1980s has greatly enhanced scientists' ability to do both types of research.

Huntington's disease research is a good example of the difficulties of forward genetics and how modern methods are helping to overcome them. The disease takes its name from George Huntington (1850–1916), a doctor from Long Island, New York. Huntington's father and grandfather worked as physicians in the same area and treated many of the same families. Just a year after graduating from medical school, Huntington noticed the pattern of a strange neurological condition that ran through several generations of one of the families. He wrote a paper describing the problem and presented it to a medical conference in Ohio. Following medical tradition, the disease is now named for the doctor who recognized it.

Huntington's is a brain disease that develops because of the premature death of particular types of neurons. Symptoms include jerky movements and a general loss of control of the body, accompanied by changes in mental abilities and behavior. The problems usually appear when patients reach their late 40s or early 50s and become progressively worse over time. As in Alzheimer's disease, the problem seems to stem from fragments of proteins that form clumpy fibers that do not dissolve.

Whereas in Alzheimer's the fibers form in the cell cytoplasm and in the space between cells, in Huntington's they mainly appear in the cell nucleus.

In 1968 Milton Wexler (1908–2007), a psychoanalyst who gained fame through his treatment of a number of prominent movie stars in Hollywood, launched a program called the Hereditary Disease Foundation when he learned that his wife, Leonore, had been diagnosed with Huntington's disease. It had already killed her father, and three of her brothers also had the disorder. Wexler knew that his two daughters had a 50 percent chance of inheriting the disease, but since the gene had not been found, there was no way to determine whether they would be affected. Wexler sponsored a series of workshops to inspire scientists to look for the cause and the cure. His daughter Nancy Wexler, a psychologist, was a member of the team who helped narrow the search down to a specific region of the fourth human chromosome.

At the time, finding the location of a disease gene was very difficult. The basic strategy was linkage analysis (described in chapter 1, "Discovering and Mapping Genes"). This involved looking for DNA sequences that were shared only by victims of the disease. The first gene maps focused on characteristics such as eye color and sex in flies that were inherited together. James Gusella and Joseph Martin of Harvard Medical School, working with Nancy Wexler and researchers from several other labs, took a slightly different approach. They took DNA samples from family members, allowed the molecules to be chewed up by enzymes, and then compared the lengths of the fragments. The enzymes cut DNA at specific spellings in the sequence, as if someone were to cut up sentences everywhere they found a certain combination of letters. Here is an example of what happens if a cut is made at the combination "re":

We are ready to leave on the retreat

would produce the fragments:

We ar   e r   eady to leave on the r   etr   eat

The allele for Huntington's disease would change the spelling or some other part of the sentence, as if one were to start with the sentence:

We are not ready to leave on the retreat

We ar   e not r   eady to leave on the r   etr   eat

All the fragments are the same except one, and it contains the letters that have been changed. In the same way, studying DNA fragments ought to show the scientists the locations of the Huntington mutation.

By 1983, scientists comparing fragments of those family members with Huntington's disease and those without discovered a difference in a region of human chromosome 4. This made it very likely that the disease gene was located nearby. Unfortunately the region was large and likely contained many genes. Wexler then pulled together a group of over 50 researchers from nine different institutes around the world to close in further. The effort involved hundreds of patients and their families. It took the consortium a full 10 years of intensive work to pin down the gene, which they named huntingtin.

More recent methods of finding disease genes draw on information in the human genome and the DNA fingerprinting techniques introduced at the beginning of chapter 3. Since the development of this method, minisatellites have provided a new way to compare the DNA of diseased and healthy members of families. When a person inherits a flawed version of a gene, it is almost always inherited along with a larger block of DNA. That block likely includes a minisatellite region with the repeats that permit DNA fingerprinting. So if a scientist can find a fingerprint inherited only by people with a particular genetic disease, the gene responsible for it is probably nearby.

Pinpointing the gene gives scientists a way to test other family members to discover whether they also have a disease-causing allele. This is helpful when a treatment is available, or when diet or other factors might slow down the course of a disease. It may also be desirable as a way to prepare people for the news that they are likely to become ill. But in cases where nothing can really be done, many doctors and genetic counselors are often hesitant to conduct certain kinds of tests unless there is another good reason to do so (see "Diagnosis and Genetic Counseling" later in this chapter).

An infectious disease can usually be cured by destroying the bacterium or virus that causes it. That is not an option in a genetic disease, where the problem is faulty information usually stored in the DNA of every cell in a person's body. There is no way—so far—to remove that bit of code and replace it with a healthy form of the information, although scientists are trying to develop methods to do so (see "Learning to Repair Defective Genes" later in this chapter). Until this can be done, the best way to attack a disease is to discover as much as possible about why a particular mutation causes its symptoms. That means learning how it affects the chemistry and behavior of cells.

Molecules have a wide variety of functions in cells that may be disturbed in disease. Some mutations damage a gene so that it cannot be used to make a protein, like the frameshifts discussed in chapter 1. Others slightly alter the protein's shape and chemistry so that it is unable to interact properly with other molecules, blocking the transmission of information through the cell, the activation of important genes, or other functions. The cell may no longer respond properly to stimuli that tell it when to divide or how to specialize. The mutation in Huntington's disease creates copies of the huntingtin protein that are too long. This changes the chemistry of the protein so that it sticks together in large clumps that cannot dissolve. Eventually this leads to the death of cells that are essential to the functions of the brain.

Some genetic flaws do not directly affect proteins; scientists are just beginning to understand how they may lead to disease. For 15 years Friedrich Luft, a physician who specializes in kidney diseases at the Max Delbrück Center for Molecular Medicine, in Berlin, Germany, has been trying to track down the cause of a rare genetic defect that causes unusually high blood pressure and a less serious symptom: People who inherit it develop unusually short fingers and toes (a condition called brachydactyly). The effect on blood pressure is so serious that it usually leads to death through strokes before the age of 50. Luft and his colleagues discovered that the "mechanical" cause is a problem with a blood vessel in the brain; it is sharply twisted, like a garden hose with a kink in it. Just as twisting the hose causes a buildup of pressure, the twisted vessel raises pressure in the circulatory system. During their research Luft's lab turned up several families across the globe with different mutations that cause similar problems. By comparing the DNA fingerprints of healthy and unhealthy members of each family, the team narrowed down their search to a specific region of chromosome 12. Strangely, the region does not seem to contain any genes. Luft hypothesizes that it might instead encode an RNA whose function is to block the activity of other molecules. But for the moment, the process that links the genome to the disease is a mystery.

MULTIFACTORIAL DISEASES

The search for the causes of Huntington's and hypertension show how difficult it has been to pinpoint specific disease-related genes—like looking for a particular needle in a haystack. Nevertheless, scientists have now uncovered thousands of Mendelian diseases, each caused by single changes in the genetic code. Most are very rare: Altogether about 5 percent of the population is affected by one of these known conditions. Many more diseases are thought to arise when people inherit combinations of certain forms of genes. Such *multifactorial diseases* include forms of cancer, asthma, Alzheimer's disease, diabetes, multiple sclerosis, schizophrenia, and many more. Finding the forms of genes that are responsible is much more difficult—like searching haystacks made entirely of needles, and having to find two or three that match.

Part of the difficulty in the process of discovery lies with the fact that humans—as opposed to peas, or laboratory mice—usually have a small number of offspring. This makes it difficult to detect statistical trends, which is why election polls and other studies of complex behavior or health issues depend on data from large, representative groups. A related problem is that the search requires comparing healthy people with those who have a certain disease, which is not always easy to determine. Symptoms strike harder in some people than others—a phenomenon called *penetrance*—or appear much later. They may also depend on additional factors such as a person's diet, whether he or she smokes, or other aspects of the environment.

Distinguishing a person with the condition from those who do not have it is a crucial step in charting the family pedigree, which is necessary before linking a disease to a region of the genome. Nancy Wexler and her colleagues needed several large families from Venezuela and elsewhere to find huntingtin, even though it is caused by a single gene. Finding multifactorial diseases will require much larger groups.

In the 1990s, a group of researchers in Iceland thought that their country might provide a head start on the problem. The population of the country is about 290,000, with characteristics that would seem to make it ideal for genetic studies:

- Since its settlement in the ninth century, the country's population has been isolated to an unusual degree. Few Icelanders have married and had children with foreigners, so there has been a very low influx of foreign genes. Homogenous populations make the best subjects in the search for the causes of genetic disorders, because disease-causing alleles stand out better against a uniform background.
- There are extremely accurate records of births and family trees stretching back nearly 1,000 years, making it easy to construct pedigrees of people's relationships over long periods of time.
- Thanks to a national health system, detailed medical records of the population have been kept for several decades.

The situation seemed promising enough that in 1996 scientists founded the company deCODE Genetics, which intended to harvest the Icelandic information to find diseases caused by combinations of genes.

The project needed extensive medical records from the population, so deCODE proposed the creation of a national Health Sector Database that would collect information from the medical and genealogical records of all Icelanders. The plan raised serious ethical concerns. Normally the law requires consent from every person who participates in medical studies, which would clearly be impossible in this case. The Icelandic legislature passed a bill in 1998 to permit the project to go ahead without citizens' consent, but five years later the country's supreme court ruled that the database violated the citizen's rights. deCODE plans to continue to work on the same scientific questions, but with a different approach. The company strategy has also changed with the discovery that the "Icelandic genome" has not been as isolated from external influences as was believed.

Despite these setbacks, however, the company has used data from Icelandic patients who gave their consent and groups from elsewhere in the world to identify forms of some genes connected to cancer, schizophrenia, and cardiovascular disease. In September 2008, deCODE reported the discovery of two gene variants that increase people's risk of developing bladder cancer. Working with thousands of patients and control subjects from Iceland and the Netherlands, the company identified a DNA sequence on chromosome 8 that increased risk of the disease by 50 percent in people with two copies of the sequence. About 20 percent of people of European descent carry two such copies, making bladder cancer the sixth most common type of cancer in the United States. The study identified another sequence, on chromosome 3, which increases the risk of the disease by 40 percent among people who inherit two copies.

Kári Stefánsson, an Icelandic geneticist and the chief executive officer of deCODE, says that the discovery of the genes should help doctors diagnose and treat patients. "In all cancers, the ability to identify individuals at high risk, screening them intensively and intervening early, is the key to improving prevention and outcomes," he says. The company offers a service to analyze the genomes of private customers, who obtain samples of their DNA by running a cotton swab along the inside of their cheek and send it in. Samples are checked for alleles linked to over 30 common diseases and deCODE supplies customers with an online analysis of any risk factors that are discovered. It also provides information about a person's historical ancestry, using the same techniques that allowed scientists to identify Mitochondrial Eve and Y-chromosomal Adam (described in chapter 2).

In the meantime, similar efforts are underway to recruit huge numbers of people in the United States, England, the Netherlands, Germany, and many other countries to carry out searches for multifactorial diseases.

Nearly all physicians agree that discovering connections between genes and diseases is important and useful. But they are concerned about how information will be used (see the next section, "Diagnosis and Genetic Counseling"). Genetic tests that could accurately predict the development of specific diseases or problems could be helpful (or misused), but even in the most straightforward Mendelian diseases, environmental and other genetic factors can influence the penetrance of a genetic problem. Additionally, studying a person's genes also reveals things about his or her parents' genes—things they might not want to know. If a young child is diagnosed with a mutation in the huntingtin gene, it means that one of the parents carries the mutation and siblings might as well—and the diagnosis might come before symptoms have appeared and anyone suspects that the family is affected. Additionally, a genetic test of an embryo may influence a family's decision to terminate a pregnancy. These issues are explored more deeply in the next sections.

DIAGNOSIS AND GENETIC COUNSELING

In September 2008 Christina Applegate, an actress best known for her role in the television series *Married with Children* and *Samantha Who?*, publicly announced that she had undergone a double mastectomy to rid herself of breast cancer. Doctors had already found and excised a small tumor from one breast.

They believed they had removed all of the cancerous tissue, but Applegate was worried that the disease might return. She had been vigilant about her health and had undergone regular tests because her own mother was a victim of breast cancer who had been through two years of chemotherapy and eight surgeries. Applegate suspected that she might have inherited a form of a gene known to increase women's chances of developing breast cancer, so she underwent a genetic test. Her doctors discovered that she carries a form of a gene called BRCA-1 known to put women at high risk for the disease. (The connection was found by Mary-Claire King, whose work is described later in the chapter.) Women with the defect have a 90 percent chance of experiencing breast cancer at some point during their lifetimes, compared to a 12 to 13 percent risk in the general female population.

Potentially, each new discovery of a connection between a gene and a disease can be turned into a tool for diagnosis. It is now possible to extract DNA from a cell, fetus, child, or adult and examine the sequence for known problems. The medical professionals who carry out such tests and interpret their results are called genetic counselors. The purpose of a test may be to explain a disease whose symptoms have begun to appear or catch a problem before it arises. Increasingly, genetic tests are being performed on fetuses to discover conditions that may lead to severe health problems or death. By 2008 more than 1,200 tests were available. They are classified into several types: prenatal, newborn, diagnostic (usually to try to understand problems that have already appeared in a patient), predictive (to see if a person is likely to develop a disease), and forensic testing.

Most counselors and members of the medical community agree that three conditions should be met before a test is administered:

- the chance is greater than 10 percent that a person has a particular mutation;
- a counselor must be on hand to explain the pros and cons of testing and interpret the results;
- the results of the test will influence how the patient is treated.

In Applegate's case, the incidence of breast cancer in her family and her own tumor made the test advisable—particularly since it might provide valuable information that could help her work out a treatment plan with her doctors. In other cases, knowing a risk factor can help a person make good choices about diet, exercise, or other behavior that might prevent additional risks or slow the onset of a disease.

One of the greatest concerns about genetic tests in general is that parents might use them to investigate their unborn fetuses and choose the types of children they want to have—leading to a modern form of *eugenics*. This term refers to the idea that the human species might be somehow improved by allowing certain people to reproduce or preventing others from doing so—or by allowing certain babies to live and others to die. If this were done on a large scale, and everyone sought to promote or eradicate the same alleles, it might slightly change the frequency of particular alleles found in the next generation. There have been attempts to do so: In the early 20th century a number of prominent doctors and scientists promoted eugenics practices through programs that were based on a very serious misunderstanding of human genetics. (The misconceptions became clear in the 1960s, through the work of Luigi Luca Cavalli-Sforza and other scientific investigations of human populations and race, described in chapter 3). The programs led to the forced sterilization of thousands of people with mental and physical handicaps in the United States, and eugenics ideas were used by the Nazis to justify the Holocaust.

This dark chapter in the misuse of science has had a significant impact on current practices in genetic counseling. Today the main reason to conduct a prenatal genetic test is to detect health problems that can be treated in the child—in some cases, even while it is in the womb.

There are three general types of prenatal tests: noninvasive, which examine a fetus with methods such as ultrasound or listening to its heartbeat; somewhat invasive methods, such as testing the mother's blood for the presence of proteins that might indicate problems; and invasive methods such as amniocentesis, which extract fluid containing embryonic cells from

the mother's amniotic sac. The choice of a method is important. Noninvasive procedures result in a higher rate of "false
positives," which means that the test suggests the presence of a
problem that is not really there. Invasive procedures are statisti-
cally more accurate, but this has to be balanced against a small
risk of harming the mother or child.

Even when confronted with an accurate diagnosis of a severe problem in an embryo, many women choose to carry an
embryo full term. As they participate in the birth and growth of
the child, many of them claim that it is an uplifting, life-chang-
ing experience.

Genetic testing raises additional ethical concerns, such as
the fear that a diagnosis will not be interpreted correctly. Many
people seem to believe that genes control their fate, although
the presence of a particular allele is rarely accompanied by 100-
percent penetrance. A person might avoid some symptoms
through diet or early treatment, and by the time they appear,
new treatments might be available. Yet even in the absence of
very reliable tests, insurance companies and employers have
already expressed an interest in gaining access to the results.
Insurance companies make profits through a careful analysis
of the probability that their customers will become sick or die.
Businesses hope to employ people who will remain healthy.
While this is understandable from a business point of view, doctors, patients, and medical ethicists almost universally regard
the release of such data as an invasion of privacy and a viola-
tion of the practice of doctor-patient confidentiality. The same
concerns brought to a halt the practice of giving data from Icelandic citizens to the company deCODE, described earlier in the
chapter.

The U.S. government has taken steps to ensure that genetic
information will not lead to discrimination in the workplace. On
a Web site devoted to "Genetics Legislation," the Department
of Energy announced the passage of the Genetic Information
Nondiscrimination Act (GINA), which was signed into law by
President George W. Bush on May 21, 2008. The act "prohibits
U.S. insurance companies and employers from discriminating
on the basis of information derived from genetic tests. . . . It for-

bids insurance companies from discriminating through reduced coverage or pricing and prohibits employers from making adverse employment decisions based on a person's genetic code. In addition, insurers and employers are not allowed under the law to request or demand a genetic test." The law follows an executive order enacted by President Bill Clinton in 2000 that imposed similar rules on federal departments and agencies.

The earlier measures, however, only affected the federal government's practices and left the legislation of the rest of the workplace to the states. Thus in 2001 the American Management Association reported that ". . . nearly two-thirds of major U.S. companies require medical examinations of new hires. In addition, 14% conduct tests for susceptibility to workplace hazards, 3% for breast and colon cancer, and 1% for sickle cell anemia; 10% collect information about family medical history." Collecting such information did not necessarily mean that it would influence decisions about hirings or promotions, but the potential was there. The new law should eliminate any abuses.

LEARNING TO REPAIR DEFECTIVE GENES

Today doctors are usually limited to treating the symptoms of genetic diseases rather than their underlying causes. Although Friedrich Luft and his colleagues do not yet understand exactly how a defect in chromosome 12 causes high blood pressure and brachydactyly (see "Finding Disease Genes" earlier in this chapter), they have been able to keep patients alive by giving them antihypertensive drugs. Another strategy is to find an inhibitor—a drug that blocks the negative effects of a misbuilt protein. But researchers have greater ambitions, dreaming of a day when they will be able to directly repair damaged genes. There will probably never be a "one size fits all" treatment for these disorders. Different approaches are needed to treat inherited genetic defects, such as the gene responsible for Huntington's disease, and somatic mutations that arise over a person's lifetime, which are frequently the cause of cancer. The main difference lies in the fact that with inborn defects, a person's immune system is adjusted to the defective molecule and cannot recognize it as something unhealthy. With somatic mutations, however, something has changed, and it might be possible to train the immune system to respond to the problem as if the body is being attacked by an infectious agent.

"Gene therapies" aimed at both types of problems are currently under development; some have advanced to the point that they can be used on human patients in clinical trials. The list below summarizes some of the approaches that are being tried.

- *Viral therapy.* The aim is to take relatively harmless viruses and adapt them to deliver healthy versions of genes to cells. Viruses hijack cells by bringing in unhealthy genes, RNAs, and proteins, often evading an immune response. Researchers would like to use them to accomplish the same thing with healthy molecules. Some viruses (such as HIV, which causes AIDS) attack only specific kinds of cells, a characteristic that therapies hope to imitate—a virus carrying healthy molecules should infect only damaged cells. Current therapies start with a virus that is as harmless as possible in the first place, such as the adenovirus. Genetic engineers remove the molecules that allow it to reproduce so that it will not spread. They add molecules that tell it what type of cell to infect, and pack it with healthy DNA or RNA. Before these therapies can be widely used, scientists must be certain that the viruses will not undergo mutations that cause them to spread to other tissues, or to be transmitted to other people. This strategy can be used both with inherited and somatic mutations.
- *Therapeutic cloning.* Here the goal is to remove cells from a person's body, change them to recognize diseased cells through genetic engineering, and return them to the body. Researchers usually start with immune system cells (such as white blood cells called T cells) and outfit them with receptor proteins that recognize *antigens* (foreign molecules). In this case the antigens would be de-

fective proteins on tumors or other diseased cells rather than molecules on a virus or a parasite. Using the strategy requires identifying a molecule that uniquely appears on cells that need to be destroyed. The modified T cells would then stimulate the immune system to respond. Another idea is to take stem cells, repair their defective genes, and return them to the body, where they will hopefully reproduce and replace cells damaged in conditions such as Huntington's disease or heart disease.

- *DNA vaccines.* Traditional vaccines train the immune system to recognize viruses or parasites by injecting dead pathogens, weakened strains of pathogens, or their molecules into the bloodstream. Immune system cells recognize them and mass-produce antibodies against their molecules. If the severe form of the pathogen now cause an infection, the body is prepared; the antibodies dock onto copies of the invader and signal their destruction. The same thing happens without a vaccine, but the body's response may be too slow to prevent the development of serious symptoms or death. Most vaccines are based on the recognition of proteins; researchers are now trying to provoke immune system responses by directly injecting DNA from the pathogens. DNA offers the advantage that it is easier to mass-produce than proteins or whole viruses and is likely to have fewer side effects.

- *RNA interference* (RNAi). In the late 1980s researchers discovered that cells produce a wide range of small RNA molecules that are not used to produce proteins, but rather to block the use of *messenger RNAs* to make them. This is now being turned into a method to prevent cells from producing defective forms of proteins. Scientists can create small artificial RNAs that target a particular messenger RNA and shut it down. The main technical obstacles are delivery—getting the molecules to specific target cells—and producing RNAs that have a long life span so that patients do not have to receive constant treatment.

The clinical trials that have been carried out so far have met with both successes and failures. Before gene therapies and the other methods become standard practice in the treatment of disease, researchers will need to acquire a great deal more knowledge of the immune system and fundamental processes within cells.

THE GENETICS OF CANCER

So far this chapter has focused mainly on diseases that arise through inborn genetic defects. Many other diseases begin with somatic changes—mutations that occur during a person's lifetime, in particular cells. They happen for a variety of reasons including exposure to toxic substances, infections, or even as random mistakes that occur during the process of cell division. Most defects are either harmless or they cause such serious problems that the cell dies. Cells have safeguards to protect themselves from somatic mutations, usually in the form of mechanisms that recognize and repair some of them. But others slip through to disrupt crucial functions such as the timing of the cell's reproductive cycle or its specialization, and this may lead to cancer.

In the "textbook" view held by most scientists until recently, most types of cancer begin with one of these somatic mutations. They are particularly dangerous when they occur in stem cells, which often reproduce quickly and have not yet completely differentiated. If a generic cell follows its normal course of development, it goes through several rounds of "decision-making" (the activation of specific genes) during which it becomes more and more specialized. At the end it should have a specific shape, should be programmed to carry out its functions, and should stop dividing. If somewhere along the way this developmental program is disturbed—for example, if there is a mutation in one of the molecules that guides its specialization—the cell may reproduce too often and form a tumor. This new mass of cells disturbs nearby organs and the body's functions. The tumor may then undergo *metastasis*: Some of its cells may detach themselves, migrate to other parts of the body, and form new tumors. Metastasis is usually the step at which a benign tumor turns fatal: Once it begins, new tumors may arise almost anywhere, and it is no longer possible to remove all the cancerous cells through surgery. In the traditional view, most scientists believed that tumor cells underwent additional mutations to behave this way.

In many types of cancer, this is almost surely the case. One example that has been extensively investigated is colorectal cancer (CRC). For many years Hans Clevers's lab at the Hubrecht Laboratory of the Netherlands Institute for Developmental Biology has been studying the causes and development of CRC, which is one of the most frequent forms of human cancer. The group has found that an important developmental signal triggered by a molecule called Wnt—introduced in chapter 1 for its role in the development of skin and many organs—has a role in the formation of new tissue in the gut.

But defective Wnt signals contribute to cancer. The reason lies in the fact that the cells that line the gut confront an aggressive environment—one of their jobs is to keep substances that are toxic from entering the body—and they wear out. They are replaced by fresh cells that arise from stem cells, which reproduce in deep, well-like holes called crypts. The Clevers lab developed strains of mice with a defective Wnt pathway and proved that the signal was necessary to help the cells divide, which allowed the gut to maintain a stock of stem cells that would be needed later. Next they wondered what would happen if the cells were stimulated with too much Wnt. When they tried this in another strain of mouse, the cells became cancerous.

Usually the tumors were not dangerous, because the cells were unable to escape from the crypts. Work by Eduard Batlle and his wife Elena Sancho, two postdoctoral fellows in the lab, showed that two molecules called Ephs and Ephrins keep the cells locked in the crypts. If a second mutation occurs that affects these molecules, the cancer can move upward, grow in an uncontrolled way, and develop into a dangerous metastasis that spreads to other tissues. So the development of CRC probably requires at least two mutations: one that accelerates the division of stem cells, and another that allows them to escape.

Intestinal Crypt Cells and Cancer

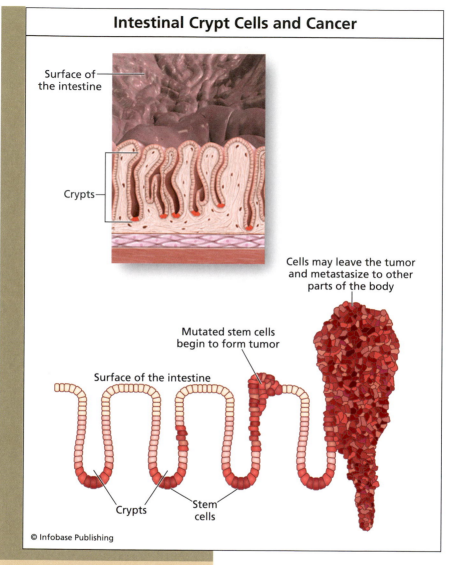

Surface of the intestine

Crypts

Cells may leave the tumor and metastasize to other parts of the body

Mutated stem cells begin to form tumor

Surface of the intestine

Crypts

Stem cells

© Infobase Publishing

Above: The lining of the intestines are interrupted by deep well-like holes called crypts. Below: Stem cells at the bottom divide and then move upward, specializing on the way, to replace worn-out cells on the surface. If these stem cells undergo mutations, they can develop into colorectal tumors—one of the most common forms of cancer. Some of these cells may leave the tumor and metastasize, building new tumors in other parts of the body.

While many cancers seem to require a series of mutations to become dangerous, some may arise through a problem with a single molecule. Such genes are known as *oncogenes* or *tumor suppressor genes*. The difference has to do with the function of the protein that the gene encodes. An example of an oncogene is Myc, a protein that binds to a large number of genes involved in the cell cycle. Mutations in Myc can make it too active, causing cells to divide too often.

BRCA-1 and BRCA-2 are other genes that can cause cancer when they become defective, but for another reason. The function of the healthy forms of these molecules is to repair damaged DNA, which can occur as a result of exposure to radiation, other types of environmental contamination, or through mistakes in cell division. Since this type of damage may cause cancer, BRCA-1 and -2 are known as tumor suppressor genes. If they are defective, the damage may slip through.

Some families carry versions of oncogenes or tumor suppressor genes that do not function properly. Researchers have mapped hundreds of different versions of BRCA-1 and -2 in human populations. Certain versions have been connected to a high risk of developing tumors, particularly breast cancer.

STEM CELLS AND CANCER

One of the most interesting areas of modern cancer research involves the connection between stem cells and cancer. This can mean two different things. One theory is that cancer arises because stem cells become defective and fail as they are supposed to differentiate into specialized daughter cells. Instead, they produce cells that reproduce too quickly, fail to develop properly, and become tumors. The other theory is that tumors arise from a more basic type of "cancer stem cell." This cell is already defective, and it divides to produce many different types of daughter tissue, including cells that are destined to metastasize and invade other parts of the body.

The first idea is a traditional one that goes back to the early days of cancer research. Theodor Boveri (1862–1915), a German cell biologist, proposed that cancer begins with a failure of cell division that divides chromosomes unequally. It was an amazing hypothesis to make at a time when scientists were only beginning to guess at the location of hereditary information in cells. Boveri conducted a number of key experiments and concluded that the information needed to create new organisms was divided up among different chromosomes. Cells had to inherit a full, high-quality set of instructions to carry out their functions. Otherwise they might develop into cancer.

Boveri made another important discovery: that a small structure in the cell called a *centrosome* usually plays a key role in properly dividing chromosomes. In the daily life of the cell, the centrosome usually sits near the nucleus and is the source of growth of a vast network of fibers called *microtubules.* These fibers act as a scaffold that gives the cell its shape and are also used as routes along which molecules are delivered to the places they are needed. Before cell division, the centrosome divides, and the two copies move to opposite sides of the nucleus. The microtubule system is rebuilt to form an elegant structure called the *mitotic spindle.* Now the fibers are used as towing lines to pull copies of the chromosomes in opposite directions. The centrosomes usually sit at the poles of the spindle. Boveri recognized that if they did not move to the right places, or if they failed in some other part of their job, daughter cells might receive the wrong number of chromosomes. This could lead to cancer.

In the meantime defects in chromosomes and their improper sorting have been confirmed as common features of cancer cells. The link to stem cells lies in the special way that they divide. Stem cells often reproduce asymmetrically: They create daughters of unequal sizes, one of which will become a new stem cell while the other specializes. Setting up this asymmetry usually requires that one centrosome is off to one side of the cell, while the other remains in the center. In spite of this, each daughter should receive a complete set of chromosomes. But if something goes wrong the centrosomes may not be properly placed, or the chromosomes may be divided unequally. This leads to the type of "genome instability" seen in Fanconi anemia (see the sidebar on Mary-Claire King) and is a hallmark of cancer. In general, this confirms Boveri's hypothesis that cancer can be caused by improper cell division, and it provides a link between the development of stem cells and cancer. Researchers are not yet sure whether genome instability and tumors are caused by the behavior of the centrosomes, or whether their disruption is due to a deeper underlying cause.

Mary-Claire King: Cancer Pioneer and Activist

The contribution that BRCA-1 makes to breast cancer was discovered in 1990 by Mary-Claire King (1946–), a professor of genetics and epidemiology at the University of California at Berkeley. At the time, the idea that genetic factors could increase a person's risk of developing cancer was controversial. It was not the first time that King's work had provoked a scientific debate, and it would certainly not be the last. Her career has taken her all over the

Mary-Claire King, professor at the University of Washington, is a pioneer in the use of DNA sequences to study human evolution, as well as in the discovery of genes linked to breast cancer and other diseases. *(The Peter and Patricia Gruber Foundation)*

world and has often mixed science and social issues.

King was a very bright young student who finished high school early and completed her bachelor's degree in mathematics at a very young age—19. She was then accepted into graduate school to study genetics, at Berkeley, but this was the era of the Vietnam War and she interrupted her studies to become active in the antiwar movement. Later she returned to Berkeley to finish her Ph.D. under Allan Wilson, who had done pioneering work involving identifying "Mitochondrial Eve" (see

(continues)

(continued)

chapter 2). Her doctoral dissertation, which she completed in 1973, involved a large-scale comparison of the protein sequences of humans and chimpanzees. It would be 30 more years before scientists obtained complete genome sequences for these two species, but her work brought her to the conclusion that there was 99 percent identity between their genes. It was a startling proposal at the time; most researchers had assumed that there were many more differences. With time, genome sequences confirmed her conclusions.

King had been interested in cancer since a childhood friend had died of the disease, and in 1974 she launched a study to find out whether there was a hereditary component to breast cancer. She began with a group of 1,000 women descended from an Eastern European ethnic group, the Ashkenazi Jews, who suffered from breast cancer at an unusually high rate. The project took 15 years, but in 1990 she successfully demonstrated that a mutation in a single gene was responsible for the increased risk of developing the disease. Again there were many skeptics. Up to that time, most researchers believed that cancer arose from interactions between multiple genes and the environment.

King went on to establish a similar role for BRCA-2, and in a study published in 2007, she summarized evidence that eight additional genes contribute to the tumors. In the paper she demonstrated that the molecules are linked to a related condition called Fanconi anemia. In this hereditary disease, genes become damaged because cells' DNA becomes unstable: Chromosomes break and their contents become scrambled, causing defects in many genes, a loss of control of cell division and differentiation, and the development of tumors.

King's research combines the computational analysis of gene sequences with experiments and epidemiological studies of families. At the same time, her experiences have brought her into contact with social and political issues, and she has combined the tools and methods of her work with these themes in a unique way.

Many of the causes she has devoted herself to have links to her personal life. Before earning her Ph.D. she had taught science in Chile. When she returned to the United States, the country underwent a military coup and its president, Salvador Allende, was assassinated. Over the next few years many opponents of the new regime were murdered or they "disappeared." Their children were sent to orphanages or adopted by childless military officers or government officials. A group of brave grandmothers tried to regain custody, but the only way to do so was to go to court and prove a family relationship. King stepped in and compared mitochondrial DNA sequences from the children and grandparents, successfully reuniting over 50 families. Later she helped to identify civilians killed by the military in a massacre in El Salvador, and has also helped to identify remains of American soldiers missing in action.

King has been very heavily involved in the Human Genome Diversity Project, launched by Luigi Luca Cavalli-Sforza (see chapter 5). This massive survey of genetic variations in human populations across the globe has become a gold mine of information about human evolution, the migrations of ancient groups across the globe, and the links between genes and disease. King's recent work includes searching for the genetic basis of some hereditary forms of deafness and looking for genes that might be giving some people a degree of resistance to the AIDS virus.

The second connection between stem cells and cancer has to do with the fact that tumors are not simply a uniform mass of undifferentiated tissue. The laboratory of Joan Massagué, a Spanish cancer researcher at the Sloan-Kettering Institute, in New York City, has analyzed breast tumors to show that they are made up of a population of diverse cells. There are two explanations: the types are spin-offs of an original cell that has undergone different types of mutations, or the original cancer cell has behaved like a stem cell. In other words, it has undergone a process of development to produce daughters of different types. The second hypothesis is supported by the fact that a tumor—like any other large tissue—needs specific types of cells to survive. For example, it needs to create an extensive network of blood vessels, or it will starve and die. And cells on the outer surface behave differently than those in the interior.

This way of thinking about tumors leads to another question: Why do some of its cells metastasize, wandering off to colonize other parts of the body? Here, too, the reason might simply be that they have undergone mutations that allow them to do so. But recently scientists have discovered evidence that from the very beginning some tumors contain cells with the potential to metastasize.

In 2008, for example, the German clinical researchers Peter Schlag and Ulrike Stein, working at the Charité Hospital and the Max Delbrück Center in Berlin, Germany, investigated tissues taken from colon cancer patients in the hospital's oncology ward. The goal was to search for differences between metastatic cells and those that remained attached to a solid tumor. In an interview with the author, Stein said, "Until recently it was difficult to define the difference between a non-metastazing cell in the primary tumor and one that would metastasize. The old view was that when a tumor had grown for awhile, there was a second event, probably another mutation, that caused one of the cells to metastasize. If that is really how things work, you can analyze patient tissues all you want and you may never find the one cell among billions of others in a tumor that will turn into a metastasis."

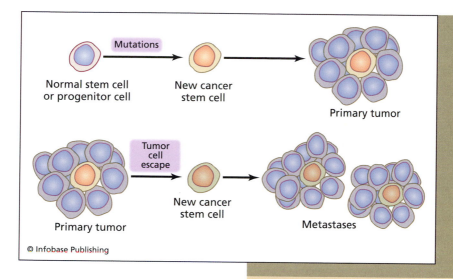

© Infobase Publishing

Two ways of thinking of "cancer stem cells": A) Healthy stem cells usually differentiate and stop dividing. Mutations might make them lose control of the cell cycle and continue to divide, forming a tumor. B) Sometimes cells become metastatic, leaving a tumor and creating new tumors elsewhere. A new mutation may be required for the cancer to spread, but in some cases the original tumor may contain cells that are already capable of metastasizing and going on to differentiate into various types of tumor cells.

But the work of Stein, Schlag, and their colleagues lends support to another scenario. "We're starting to think that there are different types of cells in the tumor from the very beginning," Stein says. "They may start as a sort of stem cell that moves into an area, replicates quickly, and then undergoes a process of development. Some of the cells it produces become migrating cells that move away. That happens over and over during normal embryonic development, and then it isn't mutations that make cells migrate. Rather it is a change in signaling and the activation and repression of genes. Now, instead of looking for random mutations, we're trying to identify specific signals and processes that become active in cells to make them metastatic."

The 2008 study revealed that a particular protein was being produced at much higher levels in metastatic cells than in cells that remained attached to a tumor. The group named the protein "Metastasis-Associated in Colon Cancer 1," or

Chernobyl and Its Impact on the Human Genome

In the wake of World War II Hermann Muller, who had discovered that exposure to radiation could cause an increased number of mutations in plants and animals, became a much sought-after expert. Governments and scientists across the world wanted to be able to estimate how severely humans might be affected by radiation from nuclear weapons or accidents involving the release of radiation. Geneticists investigated people who had lived in or near the Japanese cities of Hiroshima and Nagasaki in hopes of providing an answer. The studies did not reveal a significant jump in the number of mutations experienced by the children of exposed families. Then the terrible accident that took place on April 26, 1988, at the Chernobyl nuclear reactor facility in the Soviet Union provided another opportunity to observe the effects of radiation exposure on the human genome. DNA fingerprinting, the method developed by Alec Jeffreys (introduced at the beginning of chapter 3), has provided some disturbing insights into how families and their children are affected by radiation over the long term.

In a paper published in the April 25, 1996, edition of the journal *Nature*, Jeffreys's lab and collaborators from Russia and Belarus (near Chernobyl) presented the results of a study of 79 families who had lived in the region around the reactor continuously since the time of the accident. The following year the study was expanded to include a total of 127 families.

The scientists compared the DNA of parents and any of their children who had been born between the months of February and September 1994. Jeffreys and his colleagues picked this period because they were most interested in the effects of long-term exposure to certain types

The town of Chernobyl, in northern Ukraine, was home to a nuclear power plant that exploded in 1986, releasing large amounts of radioactivity into the environment. Today, it is a ghost town. *(Elena Filatova)*

of radiation. The parents themselves were known to suffer from a number of health problems, including an increased rate of certain types of cancer, but this study focused on whether they experienced mutations affecting egg and sperm cells that were then transmitted to children.

The researchers found that children born in Belarus carried an average of twice the number of new mutations as a group of children born in Great Britain during the same period. Although it was difficult to estimate the exact dosage of radiation that families received, there was a clear trend that "more was worse": the rate was higher among families with a higher amount of exposure. Another discovery was that the cells of men and women were equally likely to be affected.

Why were these results so different than those obtained in studies of survivors of Hiroshima and Nagasaki?

(continues)

(continued)

Jeffreys believes that part of the reason lies in the type of radiation to which people were exposed. In Japan, the most dangerous radioactive pollution faded after a short time. Things at Chernobyl were different. The accident flooded the area around Chernobyl with a massive amount of ionizing radiation from the element iodine-131, which remained at acute levels for about three months. When that threat had faded, people faced long-term levels of cesium. This substance remains in the region and is dangerous to the present day. People living there have faced continual direct exposure, and indirect exposure when the radioactive substances entered the food chain.

Another reason for the difference might lie with the way the Japanese studies were conducted. The sample of subjects was much smaller, and researchers did not exclusively focus on families in which both parents were exposed. In many of the Japanese families, only one parent had been directly affected by the atomic bombs.

The mutations that Jeffreys has studied probably have not directly affected the health of the Chernobyl children, yet other studies show that there have been major effects on the population's well-being. "A considerable increase of thyroid carcinoma in children around Chernobyl has been reported, as well as an elevated frequency of chromosome aberrations in most residents tested," the article

MACC1. Experiments in cell cultures showed that adding the molecule to cancer cells caused them to crawl away from each other. Giving it to animals with solid tumors led to the formation of metastases. Human studies show that levels of MACC1 are strongly correlated to the likelihood that a tumor in the colon will metastasize.

states. "Additionally, the frequency of congenital malformations in newborns and human embryos has increased in heavily contaminated areas of Belarus following the accident."

Jeffreys concludes that long-term exposure to smaller amounts of certain types of radiation may, in the end, cause more mutations than large, short-term doses of others. These are the first studies to show that mutations caused by chronic radiation enter the *germ line* and are passed along to the children of people who have been affected.

In 2001 the famous cancer researcher Robert Weinberg (1942–) of the Whitehead Institute for Biomedical Research and professor at the Massachusetts Institute of Technology carried out another study of victims of the Chernobyl accident. The work revealed a much higher rate of mutations in Belarus families—as much as a sevenfold increase. Jeffreys does not believe that the results are valid. He and his colleagues published a response to Weinberg's study in the same year in which they pointed out several problems with the way the study was designed and carried out.

It will probably never be possible to determine the real scope of the Chernobyl tragedy. But as the studies show, lingering radiation has had a significant effect on the residents of the region, and is likely to do so for many more years to come.

Finding markers such as MACC1 could have huge implications for patients because it may become possible to predict whether metastases will develop, and that will influence how decisions are made about treatments. Patients whose tumors reveal traces of MACC1 might need to be monitored more closely for signs that a cancer has returned or spread. In the same way, people who have inherited forms of genes such as BRCA-1 or -2 that have been linked to cancer should probably undergo more frequent tests to detect early signs of tumors.

GENES OF THE ADAPTIVE IMMUNE SYSTEM

The human body possesses two basic types of immune systems that protect it from many viruses, bacteria, and other infectious agents. The *innate immune system* is the oldest; it arose with the evolution of multicellular life and is shared by organisms as diverse as plants, fungi, insects, and other animals. The appearance of vertebrates about 500 million years ago saw the evolution of the *adaptive-immune system,* a more powerful system that is based on a very unusual set of genes. This system is so flexible that it can defend the body against a huge range of threats and stops most diseases from striking more than once. It has permitted the development of vaccines, and many researchers hope to train it to fight cancer and other health problems.

Adaptive immunity has the following characteristics:

- It can respond to a huge range of threats that cannot be foreseen;
- It must avoid attacking the healthy cells of the body it is defending;
- Cells that successfully recognize foreign proteins or other molecules make billions of identical copies of themselves—clones—to mount a full-scale attack;
- It remembers most previous infections and prevents them from successfully returning.

The main actors in this system are white blood cells called B and T cells, which come in many subtypes. Their potency comes from the fact that they can distinguish native from foreign molecules ("self" from "non-self"). This recognition system has to be flexible enough to recognize new threats; failing to detect an invader often causes serious disease. But if the system errs in the other direction and misinterprets one of the body's own proteins as foreign, the result may be a serious autoimmune disease. An adaptive immune response goes through a start-up phase in which it learns to recognize a new threat, then a phase in which it actively and broadly attacks the invader, and finally a phase when it is turned off again.

B cells are born and mature in the bone marrow, whereas T cells migrate to the thymus early in their development. Both types undergo a process of training that teaches them the difference between self and non-self. The keys to the recognition system are *antibodies* and *B cell receptors,* which are very similar—the main difference is that antibodies are secreted whereas B cell receptors remain attached to the cell.

The diagram below shows what a finished B cell receptor looks like. It is made of two identical halves that are slightly bent. Put together, they resemble a letter Y. The bottom half links the molecule to the cell membrane, and the upper half extends outward, where it will come in contact with molecules on other cells. Each half of the receptor contains two subunits. The *heavy chain* is longer, starting at the bottom, bending, and stretching up to form the two arms of the Y. Smaller *light chains* are attached to the arms.

It has taken many years to understand the process by which these proteins are made. The biggest mystery involved the fact that B cells can likely produce about 20 billion antibodies, each with a unique sequence. But the entire genome contains only about 3 billion base pairs. If each antibody were encoded in a separate gene, the genome would have to be hundreds of thousands or millions of times larger.

When researchers first began to analyze the sequences of antibodies, they discovered that the molecules had very strange characteristics. First, they found that the cells of most vertebrates are able to make two basic types of light chains and five types of heavy chains. The heavy chains have a lower section called the *constant region* that comes in only five types. But every cell produced a unique version of its upper *variable region.* The same was true of light chains, which also had constant and

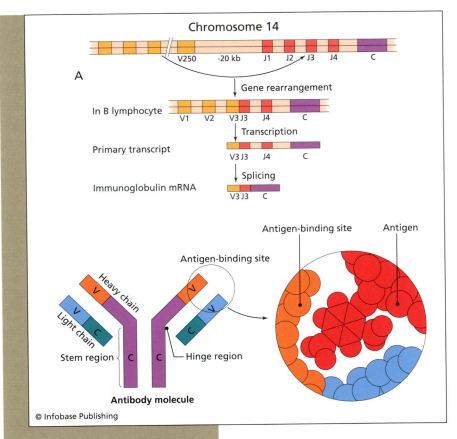

Chromosome 14

V250 -20 kb J1 J2 J3 J4 C

A

↓ Gene rearrangement

In B lymphocyte V1 V2 V3 J3 J4 C

↓ Transcription

Primary transcript V3 J3 J4 C

↓ Splicing

Immunoglobulin mRNA V3 J3 C

Antigen-binding site Antigen

Antigen-binding site

Heavy chain
V
Light chain
V C
Stem region C C Hinge region

Antibody molecule

© Infobase Publishing

Making antibodies. Human antibodies are made in a complex process that involves a random choice of V, J, and C modules, cutting out and destroying modules that are not used, and rearranging DNA to make a unique gene for each cell (A). When heavy and light chains have been translated into proteins, they are folded and bound into a Y-shaped structure. An opening at the upper tips of the "Y" provides a binding site for foreign proteins (B).

variable regions. The variable regions permit B cells to make nearly 20 billion unique antibodies.

As scientists searched for the genes containing the constant and variable regions, they were surprised to find that the pieces were scattered at large distances from one another in chromosomes. Heavy chains are assembled from four different types of genes that begin as separate units on chromosome 14. The types are called Variable (V), Diversity (D), and Joining (J) genes, plus the constant regions (C). As each B cell develops, the region of chromosome 14 containing these sequences undergoes a unique rearrangement. The process begins with the 12 D genes. One is randomly selected and *recombined* with one of the four J genes. This pair is then recombined with one of about 200 possible V genes.

Recombination is a process in which the DNA strand is cut, regions are discarded or rearranged, and what is left is reassembled in a new way. Cutting is handled by a special set of enzymes that are made only in particular types of white blood cells. These steps happen in a random way. The breaks are repaired, and when the DNA is transcribed into RNA, it interprets the pieces as single genes. This produces a heavy chain immunoglobulin RNA, which has the regions VDJC.

At the same time, the light chains are produced from sequences on the second human chromosome, which holds one type of light chain, or the 22nd chromosome, that holds the other. The process is similar to the construction of the heavy chain, except that light chains lack D regions, so when they are finished they contain the subunits VJC.

The RNAs are translated into proteins that are then joined to create an antibody or B cell receptor. Its unique recipe causes it to take on a unique shape and chemistry. Once it is completed, it undergoes tests to ensure that it does not dock onto one of the body's own proteins. If that were to happen, it might trigger an autoimmune disease. As the cells are completed, they are tested; those that bind to a human protein are destroyed.

The cells that survive have a huge variety of B cell receptors and are released to patrol the body. By chance, their shape and chemistry may allow them to bind to a foreign substance or a molecule on the surface of an infectious agent. When a B cell has such an encounter, it is stimulated to make a huge number of identical copies of its antibody. They spread through the body, glue themselves to the surface of the invader, and call up T cells and other white blood cells that destroy it. A special type of white blood cell called a memory cell retains antibodies that have been useful in combating prior diseases. If the infectious agent appears again, the production of antibodies can begin immediately.

In the first encounter with an infection the immune response takes time, and it may not work fast enough to eliminate a disease before serious damage has been done. Complex parasites such as bacteria and plasmodium, the microbe that causes malaria, have evolved clever ways of evading the immune system. Plasmodium, for example, has a life cycle involving several stages. As it reproduces, it hides inside red blood cells, where antibodies do not see it. In later stages the molecules on its surface change so quickly that by the time the immune system has mounted a response, it is no longer recognizeable.

Vaccines take advantage of the adaptive immune system by preparing it to recognize an invader before it has actually entered the body. They usually involve fragments of viruses, very similar strains, or weakened versions of the virus that cannot reproduce in cells. They bear proteins that are identical to those of the more dangerous virus, fooling the body into thinking that there has been an infection. They provoke the same type of response as a full-blown infection and stimulate the creation of memory cells. Now, if the dangerous version of the virus appears, the immune system is primed to cope with it.

One major question that researchers have had about adaptive immunity is why it is not triggered by tumors. In some ways cancer cells are similar to an infection; their surfaces often bear strange molecules (caused by mutations in genes). Somehow, though, they evade the body's immune responses. It might be possible to develop a sort of vaccine which "trains" B and T cells to perceive and attack the tumors, and laboratories across the world are trying to do so. The risk, of course, is that cancer cells occupy a strange position between "self" and "not-self," and a therapy could lead to autoimmune reactions. Yet this strategy has already been used successfully in several clinical trials. Most scientists are confident that over the next few decades, some sort of vaccine or gene therapy will become a standard tool in the fight against cancer.

PHARMACOGENETICS AND THE DEVELOPMENT OF INDIVIDUALIZED MEDICINE

Every drug sold in a pharmacy or prescribed by a doctor comes with complex and bewildering instructions. A single medication may cause "drowsiness or excitability," "constipation or diarrhea," or have other side effects that seem contradictory. Doctors have always known that people often react very differently to the same treatment, but it has usually been difficult or impossible to determine why. Since the 1950s, researchers have known that genetic factors may play an important role, and experts now estimate that they may account for anywhere between 20 and 95 percent of the variation in patients' responses to drugs. New *genomic technologies,* developed as a by-product of the human genome project, are now being used to discover some of the connections. This approach, called pharmacogenetics, has already produced some useful tools to predict how patients will respond, and it is an important step toward individualized medicine: treatment based on unique features of a person's genome.

In some cases a drug's effects are strongly dependent on a single gene. Most drugs are small molecules that have been extracted from a natural substance, purified, and "optimized" through chemical changes that increase their potency. Usually they work by docking onto a cellular protein and influencing its behavior in some way—for example, by blocking the binding site for another protein. What a drug binds to and how well it does so are determined by its shape and chemistry. It should function like an adaptor plug designed to fit into a specific socket. Here the shape of the "socket" is just as important as that of the "plug," which makes it difficult to develop generic cures; a worldwide study shows that most human proteins come in thousands of variants. (This effort, called the Human Genome Diversity Project, is described in chapter 5.) Many of the variants have slightly different structures and chemistry that could influence the activity of a drug. The clearest cases are those in which there is a variation directly in the molecule the drug is supposed to dock onto. Scientists have identified more than 25 examples of reactions based on differences in direct drug targets.

But this type of direct interaction is only the first step in a process that usually involves many molecules and ultimately determines a drug's impact. Human cells have over 30 families of molecules that are involved in transporting or processing drugs. Defects in any of them may cause an adverse reaction. That is the case for several members of one of these families, called the Cytochrome P-450 enzymes. One of them is the protein CYP2D6, whose normal function is to help break down foreign molecules and clear them from the body. Usually it activates codeine and plays a key role in the response to other painkillers. But over 75 variants of CYP2D6 have been found in humans, and some of them cannot metabolize codeine. And the protein does not have to be defective to have an influence on a drug's impact. A large number of people in East Africa have multiple copies of the gene, which means that they often do not respond to normal doses of the drug.

A test to determine a person's form of CYP2D6 may reveal whether codeine should be prescribed. Another famous case involves an enzyme called TPMT, which processes a class of drugs called thiopurines, used in treating childhood leukemia and several autoimmune diseases. When these drugs were first introduced, doctors noticed that a proportion of their patients suffered a dangerous side effect: Their bone marrow was being destroyed. Further studies revealed that one in 300 people have two variants of the TPMT gene that do not function. If they are given standard doses of the drug, they may experience severe bone marrow damage. With this knowledge, doctors can test a patient's TPMT activity to determine whether this is likely to happen.

Just as many diseases are likely linked to a combination of genetic factors, researchers believe that variations in drug responses often depend on multiple genes. Finding such links is much harder because they require a huge number of subjects, including many members of the same family taking a particular drug, and the process is extremely expensive. Here is where researchers are hoping that genomic technologies will soon help, particularly as the costs of DNA sequencing keep falling. If it were possible to decode an individual's entire genome for a reasonable price, the sequence would contain all the information necessary to scan for forms of genes that have been linked to diseases or adverse drug reactions. The alternative is to look at specific genes, but then scientists need to know which ones to investigate.

The original human genome sequencing project cost American taxpayers about 2.7 billion dollars—less than the real total price of the project, which included a number of international institutions. Currently the National Institutes of Health have funded a number of technology development initiatives to reduce prices. In the short term the NIH hopes to obtain a sequence for about $100,000, which will make it feasible to obtain hundreds or thousands of individuals' genomes. This would greatly speed up the search for disease- and drug-related genes. Eventually, the NIH hopes that a complete sequence will cost only about $1,000. At that point, it will be feasible to complete individuals' genomes and use them in diagnosing health problems as well as making decisions about treatments. A patient could carry his entire genome around on a chip card or similar device, providing access to the information whenever it was needed.

5

The Genetics of Behavior, Human Diversity, and Society

The previous chapters of this book have focused on the relationship between genes and the human phenotype: how the body develops and functions. But the genome's influence on people's lives does not stop there: It has an impact on behavior, human society, and culture. These areas are what evolutionary philosopher Richard Dawkins (1941–   ) calls "the extended phenotype." Behavioral, social, and cultural factors influence which mates people choose, how long they live, and how many children they have, so they contribute to heredity and human evolution.

The fact that genes play a role in these areas does not mean that they dictate a person's fate. The same set of genes could produce a citizen of ancient Rome or a modern American. A person's language, beliefs, tastes, and knowledge of the world are the product of culture. Yet by giving a person a brain and body with many limitations, and simultaneously the capacity for certain types of learning and social adaptations, genes help define the limitations and possibilities of culture. For example, human children have a very long infancy and are dependent on care from their parents for much longer than most other animals, and this has had a great impact on society. If they became independent more quickly, nomadic tribes throughout history might have had more children. The organization and purpose of schools would certainly be different.

The information in the genome makes humans different from other primates; it also explains both the common and unique features of individuals. This chapter discusses what can be learned about behavior and society by looking at both from a genetic and evolutionary point of view. These topics have been the subject of heated debates by politicians as well as scientists, because the role of heredity and the environment in shaping human lives has a political dimension. As decision-makers try to build a safe, healthy society for their citizens and design better health and education programs, it would be helpful to know how much influence the environment has on human development and behavior.

THE SCIENCE AND POLITICS OF HUMAN DIVERSITY

The DNA sequence obtained by the public human genome project is a mixture of DNA from several different people; no one person really has this exact sequence. It is intended as a standard version—not the "best" genome, or an "ideal," but one that can be used to discover genes, explain how they contribute to building a human body, and contrast humans with other species. It can also be used to help discover what makes individuals unique, which means analyzing how their genomes differ from the reference version and from each other.

On the average, one person's genome differs from that of another, randomly picked individual by one base per every thousand. Such "spelling changes" involving single letters of the genetic code are called *single nucleotide polymorphisms,* or SNPs (pronounced "snips"). Most of these are mutations that occurred long ago; others are more recent, because mutations happen all the time as DNA is copied. If they do not interfere with a person's health or his or her ability to reproduce, they can be passed along to children. Since every new embryo experiences a few unique mutations, SNPs can be used to make family trees or determine how people are related to each other, using the methods described in chapters 2 and 3.

One way of comparing individuals would be to sequence their entire genomes and compare all the SNPs, but this is too expensive, and it is unnecessary because the relationships can usually be established by looking at much smaller regions of the genome. These shortcuts are haplotypes (introduced in chapter 2), which are collections of SNPs, arranged in the same order on a chromosome, shared by several people. Chapter 2 described how these methods have been used by Stanford geneticist Luigi Luca Cavalli-Sforza and others to trace ancient human migrations. Another use of haplotypes is to search for genes linked to disease. Finally, haplotypes can give researchers a measure of the extent of human diversity, and answer questions such as how fast the species is evolving.

In the early 1990s Cavalli-Sforza, along with Allan Wilson and Mary-Claire King (featured in chapter 4), began an ambitious project to measure human variation by sampling DNA from populations around the globe. The idea was to conduct a sort of genetic "world census," called the Human Genome Diversity Project (HGDP). Cavalli-Sforza introduced the HGDP in a paper published in the journal *Genomics* in 1991. The aim, he stated, was to make new discoveries about the "genetic geography and history of our species." In addition, he pointed out that some human populations were disappearing, and the project might be the only way to ensure that information about their DNA would survive.

But early on the project stalled as a number of critics raised ethical concerns. In an article on the history of the project, published in *Nature Reviews* in 2005, Cavalli-Sforza summarized the main issues. "They focused especially on the fear that indigenous people might be exploited by the use of their DNA for commercial purposes ('bio-piracy'). . . . Concern that HGDP data would feed 'scientific racism' was also expressed by naïve observers."

The bio-piracy issue stems from sensitivity to a long tradition of colonialism by developed countries, particularly in the Northern Hemisphere, which has systematically exploited countries of the developing world by taking resources without returning profits to them. This has deeply affected areas such as the Amazon or Africa, which are of high interest to geneticists because of the long isolation of their populations or the historical importance of the regions. In the Amazon, mining and deforestation, along with diseases brought in by outsiders, have had a devastating impact on the survival and way of life of indigenous peoples. Nongovernmental activist organizations such as ETC (the Action Group on Erosion, Technology, and Concentration) pointed out that HGDP information might eventually bring enormous profits to pharmaceutical companies and other industries, but that the project was launched without a careful plan for profit sharing or protecting the interests of regional cultures.

Another issue has been the idea that genetic information could somehow be used in support of racist and political agendas. Until now, there has been no clear genetic definition of what it means to be African American, Anglo-Saxon, Hispanic, or any other "race"—and geneticists have been unsure that it is possible to create such a definition. In a 2006 article called "Two Questions about Race," biological anthropologist Alan Goodman, president of the American Anthropological Association, summarized the situation this way: "Americans and much of the world's population have been conditioned to think of race as a fuzzy jumble of behavior, culture, and biology: a deep and primordial mix of a bit of culture and a lot of nature."

A genetic definition of race—if it could be achieved—might give governments, employers, and others objective guidelines by which they could classify people. Some even fear that it might make possible new biological weapons that are a threat only to people with certain genetic profiles. In other words, they would target unique genes of specific ethnic groups.

This could only work, of course, if a particular group had any unique genes, and Cavalli-Sforza and most other experts believed that the HGDP would reveal that they did not. "Half a century of research into human variation has supported the opposite point of view—that there is no scientific basis for racism,"

Cavalli-Sforza writes. This had been the conclusion of every study of human variation carried out using partial information, such as the protein sequence studies carried out by Marie-Claire King, described in chapter 4. Cavalli-Sforza and most other experts believe that most fears about the HGDP are irrational and based on a misunderstanding of both genetics and science.

In spite of the scientific arguments, critics of the HGDP had a significant impact on early phases of the project, stalling it for three years as the U.S. National Research Council carried out a study on its ethical dimensions and feasibility. Finally, in 1997, the HGDP was permitted to go ahead. The project began collecting cell lines, which would be used as the source of DNA samples, from people of different ethnic backgrounds around the world. By 2002, more than 1,000 cell lines from 52 populations had been collected. The collection is currently housed at the Center for the Study of Human Polymorphism at the Jean Dausset Foundation in Paris, and is made freely available to researchers around the world.

Noah Rosenberg and members of Marcus Feldman's laboratory at the University of Southern California, in Los Angeles, were among the groups to receive cell lines for study. Their work turned up some interesting facts. Earlier studies based on much smaller sets of information had shown a surprising lack of diversity among humans compared to other species. For example, a group of western lowland gorillas living within a relatively small region of Africa showed twice as many genetic differences as the most different humans living anywhere on the globe. The findings of Rosenberg and Feldman supported one possible explanation, the hypothesis introduced in chapter 2: Modern humans have gone through a severe "bottleneck" within recent evolutionary history, meaning that the entire current human population descends from a very small group, possibly fewer than 10,000 people.

The study also confirmed geneticists' view on race: There are more genetic differences within a given population—such as the inhabitants of a particular region of Europe or the United States—than between either of these groups and a tribe living deep in the Amazon basin. "Within-population differences among individuals account for 93 to 95 percent of genetic variation; differences among major groups constitute only 3 to 5 percent," Rosenberg writes. This is smaller than earlier estimates, and it is a direct outcome of the HGDP. Rather than providing fuel to racists, the study shows that the clearest way to sort people into groups is by geographical region, rather than by superficial judgments based on skin color or other traditional ways of classifying people.

This does not mean that the differences between groups are meaningless. "Without using prior information about the origins of individuals," the study reports, "we identified six main genetic clusters, five of which correspond to major geographic regions, and subclusters that often correspond to individual populations." The information has helped answer questions about the relationships and origins of the people who have settled various parts of the globe.

One use of SNPs and other types of genome data is to detect subtle traces of natural selection. If a population is healthy, has enough to eat, and intermarries randomly, old and new alleles will be shuffled from one generation to the next in a random way. If a particular allele offers an advantage—or is harmful—it will not be passed along randomly. Over several generations, it will become more common (or more rare). These trends can be detected by mathematics. The study from Feldman's lab shows that populations that are isolated from each other have undergone drift. The longer they are isolated, the farther they drift apart.

Human adaptations were the subject of a 2006 study carried out by Eric Wang, Robert Moyzis, and their colleagues at the University of California. The scientists used SNP data that had been collected by Perlegen Sciences, a biotechnology company based in Mountain View, California. They also drew on data from the International Human Haplotype Map project (Hap-Map), another large-scale initiative to collect information about human variation.

Wang and Moyzis developed a new statistical method to examine the data and estimate what they call "linkage disequilibrium decay," or LDD. The method scans the neighborhood of DNA sequences around a particular SNP and tries to estimate the age of a particular mutation, then sees how well it has been preserved over time. The study revealed "a surprising number of alleles with the fingerprint of recent positive selection." The scientists conclude that "selection for alleles in these categories accompanied the major 'out of Africa' population expansion of humankind and/or the radical shift from hunter-gatherer to agricultural societies."

EVOLUTIONARY PSYCHOLOGY

Corals are marine animals that often live in large colonies. Some species secrete calcium carbonate, a mineral that hardens and forms huge reefs. This capacity, which is encoded in the genes of corals, allows them to create the environment in which they live. They share a reef with thousands of other species, which influences their lives and their evolution. Human society is likewise an environment that shapes its members—determining the language they speak, the clothing they wear, the food they eat, the technology they use, and the chemicals they are exposed to. These factors play a role in how long and well people live and with whom they have children, influencing heredity and the future evolution of humanity.

In the same way, the people alive today are the product of the physical and social environments of the past. Researchers believe that long ago, climate changes in Africa led to deforestation and the development of wooded plains and savannahs. This favored the selection of primates that walked on two legs, which went on to evolve into many new species, including modern humans. As early hominid species spread from Africa to the rest of the globe, they encountered new environments with different climates, plants, and animals that triggered both physical and social adaptations. At each stage, natural selection had an influence on the body and mind. The goal of the field of *evolutionary psychology* is to try to establish what factors have helped shape the human brain, behavior, and society. Few people would disagree that the behavior of ants, which have tiny

brains, is programmed by their genes. Is the same thing true of human activities like engineering, language, music, mathematics, and science, or are these types of behavior something more?

Leda Cosmides and John Tooby, codirectors of the Center for Evolutionary Psychology at the University of California at Santa Barbara, have written a "primer" for this field at their Web site (see "Further Resources"). There they explain the basic philosophy of evolutionary psychology: "The mind is a set of information-processing machines that were designed by natural selection to solve adaptive problems faced by our hunter-gatherer ancestors." Drawing on information from biology, anthropology, neurobiology, sociology, and other fields, Cosmides and Tooby hope to define the machines and identify the "genetic footprints" of natural

An artist's depiction of the life of "Rhodesian man," an interpretation of fossils found in 1921 in territory now belonging to the country Zambia. The hominid's exact place in human ancestry is unknown, but many believe it to be an ancestor of modern humans. Evolutionary psychologists hope to demonstrate how the living conditions of earlier hominids influenced the evolution of "modules" in the modern human brain. *(Amadee Forestier)* selection. This might lead to a better understanding of the human mind and how it copes with the present-day world.

The most can be learned, they believe, from skills such as verbal language. Anything so universal, the scientists argue, must have evolved from specific challenges humans faced in the past. The time frame in which they are most interested is the hunter-gatherer phase of culture, a period that comprised more than 99 percent of the history of modern humanity, and the epochs in which their primate ancestors lived. Although today's humans inhabit in a vastly different world of computers, modern transportation, and urban life, they have to cope using the mental abilities and structures developed during prehistory.

The great flexibility and range of human cultures reflect what may be humans' most important and unique characteristics: inventiveness, problem-solving skills, and a vast potential to learn. These capacities are encoded in genes and have contributed to humans' ability to survive and reproduce, and have undoubtedly been subject to natural selection.

Evolutionary psychologists think of the brain as a collection of various modules, a bit like different cards installed in computers to handle sound, graphics, and communication with other computers. Cosmides and Tooby believe that each module of the brain evolved through different pressures, rather than as an undifferentiated whole. Today interactions between the modules motivate people, help them interpret the world, and are responsible for types of thinking that are found among all humans.

The extreme version of evolutionary psychology claims that most human thoughts, feelings, and actions are specifically encoded in genes and have been selected for. Several of the studies carried out by Cosmides and Tooby suggest that humans have an innate ability to detect deception. Love also has an evolutionary value: It helps keep families together for the long periods of time that human children are dependent on their parents. But how far does the influence of genes go? Do they make a small or large contribution to whom a person loves, how well someone does in school, or whether he or she commits an act of violence?

Evolutionary scientists from Charles Darwin onward have emphasized the role that natural selection has played in shaping the human brain and body. Obviously the building plan of the human brain, its size, tissues, and the way its cells and molecules operate are the result of genes. Scientists also know that genes direct the "hard wiring" of the system; in other words, they help neurons in one part of the brain connect themselves to specific cells in other regions. Establishing the right connections is also essential to the functions of the brain and body. For example, nerves from one hemisphere of the brain have to be wired to the other so that skills managed by the two sides can be coordinated. If not, the result is a brain that has trouble coping with the daily requirements of living with other human beings. And neurons send long extensions—axons—to distant reaches of the body, so that sensations such as pain are transmitted to the brain. Other sets of nerves are linked to muscles, which gives people conscious control of their bodies. All of these processes are linked to genes, and all of them have a direct impact on behavior. If the machine does not work right, thinking and behavior cannot happen in the normal way. Mutations in specific genes have been found to affect language skills and reasoning, as well as to play a role in learning disabilities, autism, and schizophrenia. Yet few of these cases can be pinned down to single genes, and environmental factors almost always play an important role in whether they develop in a specific person.

Comparisons of humans with their close evolutionary relatives reveal striking similarities in behavior that must have a genetic basis. Jane Goodall (1934–), a famous British primatologist who has studied chimpanzees for over 45 years, discovered that the animals not only use tools but make them—which had previously been considered a uniquely human trait. She also observed community behavior, including acts of aggression and murder, that had only been witnessed among humans. These parallels reveal that the genetics of some types of social behavior stretch millions of years into the past, to a time before the divergence of chimps and humans.

But evolutionary psychology also hopes to explain the origins of behaviors that are uniquely human. Has the ability to

More clues to the origins of human social behavior have been gleaned from studies of chimpanzees, *Homo sapiens'* closest evolutionary relative, carried out by primatologist Jane Goodall and others. *(Michael Neugebauer, Carnegie Science Center)*

sing and perform music, for example, been the subject of natural selection? Or is it a side effect of having a certain type of brain structure? Critics point out that without knowing exactly what pressures have shaped the brain, it will likely be impossible to give a clear answer. Without details, they say, evolutionary psychology is little more than an attractive idea, with limited value in interpreting modern human behavior.

Stephen Rose, of the Free University of London, regards the field as simplistic because it downplays the role that other factors play in people's lives. In a personal interview with the author, he put it this way: "I don't think we can understand what it means to be human without understanding that we are evolved organisms, just as much as we are social, historical, cultural, and technological organisms. To understand human nature means that we have to understand all of these things."

Stephen Jay Gould, the famous paleontologist and evolutionary writer, criticized the field as telling "just-so stories" because there is no way to test hypothetical causes and effects in ancient history. In 1979 Gould and the population geneticist Richard Lewontin introduced the term *spandrel* as an alternative to explain how some human mental abilities might have evolved. They borrowed the word from architecture, where it is used to describe rounded triangular spaces that are left over where an arch is fit into a square frame, or where a dome is mounted on one. These spaces have no real function; they are simply by-products left over when the two shapes are fit together, and Gould and Lewontin use it to bring up a similar situation in evolution. "Evolutionary biology needs such an explicit term for features arising as by-products, rather than adaptations," Gould wrote. The brain did not evolve in order to use language, do mathematics, or make music; to do those things the brain already had to have a certain structure, which arose for other reasons through natural selection. What evolutionary psychologists call modules might have been spandrels that have not yet been selected for any specific contribution to human survival or reproduction.

Science can show the mechanisms by which organisms have evolved (by discovering mutations in genes) and when (using the molecular clock strategy described in chapter 2). These strategies require comparing different species, and allow researchers to track the origins of genes related to brain structure and study the changes they have undergone over time. But discovering that natural selection of a trait has happened does not allow a scientist to point to specific events in the environment that gave some individuals a reproductive advantage over others.

In some cases the connection between features and survival are fairly obvious—as in the case of camouflage among insects, which hides them from predators. In other cases it is possible to make elegant hypotheses. The Madagascar star orchid has a floral tube that is 12 inches (30 cm) in length, and for the plant to produce seeds, pollen has to be deposited in its depths. When a collector sent Charles Darwin the plant, he hypothesized that somewhere there had to exist a moth with a 12-inch (30-cm) long feeding tube. The idea was ridiculed, but four decades later, scientists discovered this insect. How did this extreme case of coevolution get started? One hypothesis, proposed by botanist Lutz Wasserthal in 1997, gives responsibility to the moth: a longer feeding tube helped certain insects survive because spiders occasionally hid in the flower and could eat moths that came too close. But a competing hypothesis suggests that the structure of the plant is responsible: to transfer pollen, the moth has to bump its head on one part of the orchid and reach into the tube with its feeding tube. The tongue had to be a certain length to do both. As mutations gave the plants a new shape, they could be pollinated as long as there was an insect capable of doing so.

Over time, careful study of the plant and moth may reveal that one of these two hypotheses is more feasible than the other. But the real answer lies in history, and in millions of encounters between moths, plants, and possibly spiders, so it will likely never be possible to establish a direct cause-and-effect relationship. In the same way, it may be possible to create plausible scenarios to explain the origins of language and other human abilities. But it may never be possible to identify specific pressures that encouraged their development. So the extent to which evolution can explain details of human psychology is still an open question at the beginning of the 21st century.

Yet even if its speculations cannot be proven, an evolutionary perspective on human psychology may be useful and enlightening. The epidemics of obesity and type 2 diabetes that are running rampant through the developed world may be a good example. Both conditions are closely linked to diet and exercise—two aspects of life that have changed enormously over the past few decades, which is less than a blink of the eye in evolutionary time. Today's habits are a dramatic departure from the lifestyles during which human evolution was shaped. Tastes for particular types of food might have motivated early ancestors on the African savannah to seek vital elements of their diet that they otherwise would have ignored. Early hominids needed high-energy foods like sugar and salts and fats, which all were probably difficult to get. Natural selection probably would

© Infobase Publishing

Coevolution: the orchid and the hawk moth. After seeing the Madagascar orchid, which has a 12-inch (30-cm) floral tube, Charles Darwin predicted that there had to be an insect with a 12-inch (30-cm) tongue able to pollinate it. Later scientists discovered just such an insect: the hawk moth. The length of the flower and the moth's tongue are a clear case of coevolution.

favor people with a craving for them, because they were willing to make an extra effort to obtain them (for example, by risking being stung when gathering honey). Those ancient taste preferences have survived, but in today's society they may lead to excesses that cause obesity, diabetes, and other health problems. Learning to help people with these problems may depend on recognizing that their behavior is motivated by at least hundreds of thousands of years of evolution, during which time such behavior would have been healthy and contributed to the survival of the species.

TWIN STUDIES AND THE GENETICS OF IQ

Identical twins (also called "monozygotic twins") have the same genome. This unique situation has led researchers to hope that studying their biology and behavior will give insights into some of the questions raised by evolutionary psychology and other aspects of the relationship between genes and behavior. Twin studies are being used to look for genes that contribute to mental conditions such as schizophrenia and autism, to discover the degree to which intelligence depends on genetic versus environmental factors, and to search for the causes of obesity and a wide range of other health problems.

Any parent of more than one child knows that even children raised in the same household develop distinct personalities, habits, and abilities. This would suggest that differences in genes are responsible. Things are not, of course, that simple; the environment is never exactly the same for two people. For one thing, older children may be around. And while on the average any two siblings share half their alleles, each child is a unique mix of genes from the two parents. This complexity of the genome and the environment makes it virtually impossible to distinguish genetic and environmental factors that influence an individual's development.

Identical twins, on the other hand, seem as though they would provide a unique opportunity to peel these factors apart. Even twins that have been raised together become unique individuals, which shows that the genome can develop in subtly different ways even if there are many constants in the environment. Identical twins that have been raised apart offer another perspective on the problem. If the environment were responsible for everything about a person's psychological development and behavior, then such twins should be no more similar than any other two people of the same age. However, if studies show that they share traits at a higher rate than normal, the common features may be due to genes.

A pioneer in this field was the prominent British psychologist Cyril Burt (1883–1971), who published several studies on twins between 1943 and 1966. His main interest was IQ, and over the years Burt found pairs of identical twins who had been raised apart and gave them intelligence tests. Using statistical analysis, he estimated correlation coefficients of the intelligence of the twins. Such coefficients are measured on a scale from zero to one, where a higher score means that two people are more similar; a score of 1.0 would mean that they were perfectly correlated—that their IQ scores were identical. A low score would mean that there was only a random association between two things. Usually a score above 0.8 is considered strong; anything below 0.5 is weak.

In 1943, using 15 pairs of twins, Burt discovered an intelligence coefficient of .770; in 1955 he expanded the study to 21 pairs and found a coefficient of .771. The largest study, completed in 1966, reported a score of .771 for 53 pairs of twins. Burt concluded that in the development of intelligence, heredity played a much stronger role than the environment. In the political and social environment of the 1960s and 1970s, the results were greeted enthusiastically by many politicians, educators, and others who were being forced to integrate schools and make other social changes. If intelligence were much more strongly influenced by heredity than environmental factors, then education reforms might not change things as much as people hoped.

Within a few years of Burt's death, several psychologists began to doubt the validity of his work. A Princeton psychologist named Leon Kamin found it strange that the correlation coefficients of the studies were so close to each other, despite the fact that different numbers of twins were used. In his book *The Science and Politics of IQ,* published in 1974, Kamin emphasized the cultural bias of IQ tests and pointed out some problems with Burt's studies, without directly stating that Burt had engaged in fraud. That charge came two years later, when Oliver Gillie, a medical correspondant for the *London Sunday Times,* reviewed Burt's work and became suspicious. Other researchers, including Burt's biographer Leslie Hearnshaw, quickly found more discrepancies. Based on indirect evidence, they believed that Burt had made up two of the authors cited on the original papers. (This unusual suspicion was based on the fact that the two scientists had "disappeared"—they had stopped publishing scientific papers. Later, Burt's colleagues stated that the two scientists had indeed existed.) Another issue was the number of twins in the studies: Identical twins are rarely raised in complete isolation from each other, and it was difficult to believe that Burt had managed to find 53 pairs. After a careful examination of Burt's private papers, Hearnshaw (originally a supporter of the psychologist's reputation) also concluded that his subject had engaged in scientific fraud.

The case is not closed; a more recent examination of the evidence suggests that there may be innocent explanations for what the critics have found. For example, records on the 53 pairs of twins may have been among data that was lost during World War II. But because it is impossible to reconstruct all of the data or repeat the studies, a cloud of suspicion continues to hang over Burt's work.

The fight over Burt's reputation is a side-effect of a much larger social issue because the heredity of intelligence has become a chess piece in political debates. Some politicians have come down on the side of "genes determine behavior," claiming that social programs will not significantly help to change children's intelligence, their tendency toward violence, or other types of behavior. Others have generally assigned a stronger role to the environment, suggesting that social welfare programs or projects like the racial integration of schools may have a significant impact on children's intellectual and social development. For completely different reasons, the communist regime of the Soviet Union promoted an aggressive anti-genetic philosophy in which virtually everything about an organism was assumed to have come from the environment. In promoting their new vision of society, communists believed that the economic system of a society could completely control human nature, eliminating characteristics like greed—even influencing people's genes. Their theories, however, were based on a complete misunderstanding of genetics and the interplay of genes and the environment.

Putting political considerations aside, what has been learned about the genetics of intelligence? Despite the questions surrounding Burt's studies, and despite important debates about the bias and meaning of IQ tests, most research on twins has consistently shown a strong link between IQ and genes. A 1995 review of the field by the American Psychological Association estimated that by the time twins reach late adolescence, there is a 75 percent chance of their having very similar IQs, regardless of their upbringing; this means that by the late teen years, heredity has played a major role in the level of intelligence a person has achieved. However, further research has shown that these statistics are biased toward twins and children raised in wealthy families (who are more likely to participate in studies). Twins raised in a lower socioeconomic background are likely to be more different.

The 1995 study was motivated by controversy surrounding a book called *The Bell Curve,* written by psychologist Richard Herrnstein of Harvard and political scientist Charles Murray of the American Enterprise Institute. The book correlates people's scores on IQ tests to various aspects of their lives. One conclusion is that there is a strong connection between measures of a person's intelligence and socioeconomic status. Using measurements such as level of poverty, level of education, divorce rates, unemployment, and other factors, Herrnstein and Murray discovered that a person's IQ was highly correlated with "success." In other words, a person with high intelligence was more likely to pursue higher education, stay employed, and have a stable family life than someone with a lower IQ.

The authors concluded that intelligence was largely responsible for families' social and economic problems, rather than the other way around. This was a dramatic claim, but the real controversy started around their interpretation that intelligence was hereditary and linked to race. To put their position most simply: Since people of different races are born with different levels of intelligence, it is no wonder that they achieve different standards of living in society. Giving people welfare or improving the education system would not make much difference, and one of Herrnstein and Murray's conclusions is that the government should cut social programs.

This sparked an outrage among scientists who felt that there were other interpretations of the data and that Herrnstein and Murray had picked the most dangerous one. The review by the American Psychological Association stated that there was no evidence that differences in scores on intelligence tests by members of different races had a genetic basis. Many critics pointed out that standardized test scores were averages and did not predict anything about individual performance. It was stretching things too far, for instance, to claim that education programs would not help specific students. And the study might disguise hidden environmental factors that were difficult to determine.

This is one conclusion of follow-up work by Erik Turkheimer, professor of psychology at the University of Virginia. In 2003 Turkheimer and his colleagues reported findings from a study of families of different social backgrounds. Twins that had grown up in wealthy families, they discovered, had very similar scores—indicating that in an environment of "opportunity," genetics seems to play an important role in the level of intelligence that a person achieves. When raised in poorer families, however, identical twins' scores varied widely. They were just as different as those of fraternal twins. Turkheimer concludes that in lower-income families, the environment plays a stronger role in how an individual's intelligence develops.

THE GENETICS OF COGNITIVE DISORDERS

One of the problems of research into intelligence and genes is that IQ is a very abstract way of measuring mental abilities, which are complex and hard to define. Another way to study the brain is to look at people whose abilities are impaired. The approach has been used for many years to catch a glimpse of overall brain function. Pathologists and neurologists have long had an interest in patients with strokes or other types of brain injuries, matching specific regions of the brain that have been damaged to the language or other types of skills that have been lost. This work has been modernized through the use of techniques such as functional magnetic resonance imaging (fMRI), which can track changes in the flow of blood in the living brain as a person performs different tasks. One use is to compare differences in brain activity between healthy subjects, those with injuries, and in people with impairments due to genetic defects. Combining this with other methods is steadily providing new insights into the relationship between genes, major brain structures, and how the organ functions during different types of mental activity.

Among the discoveries are genes that seem to play a role in conditions such as schizophrenia, *autism,* and bipolar disorders, and some types of mental retardation, as well as problems with language, mathematics, and other cognitive abilities. This is a vast field that will only be briefly introduced here. Interested readers can explore the topics in detail at the Web site of the "Genes to Cognition" project (www.g2conline.org) created by the Dolan DNA Learning Center at Cold Spring Harbor Laboratory, New York (see "Further Resources").

One molecule associated with mental retardation and other disruptions of brain function is a gene called "fragile X mental retardation 1," or FMR1. While it is used to create proteins by cells throughout the body, it seems particularly important in the brain, where it is thought to help establish connections between nerve cells. The link between a gene and retardation was first noticed in the 1940s by two Irish researchers working in London—James Purdon Martin at the National Hospital and Julia Bell at the University College. They carried out a study of an extensive family that had 11 males who were clearly mentally disabled and several mildly affected females, concluding from the pattern of inheritance that the condition was linked to a defect on the X chromosome. During the 1960s and 1970s researchers discovered that people who inherited the disabilities had an X chromosome that seemed to be broken at the end.

In 1991 an international team of researchers pinpointed the condition to a gene in this region of the chromosome, which they named FMR1, and began to understand what type of defect was involved. Like a number of other genes, FMR1 contains a long section of small "repeats": a small DNA sequence that appears over and over, like a record whose needle has gotten stuck. Repeats in genes often consist of just three letters. In FMR1 they are composed of the bases "CGG." In most people, the gene has about 30 copies of this pattern, although some have as few as six copies. In families with fragile X syndrome, however, the gene may contain 200 copies of the sequence—and in some cases even more. A male who inherits more than 200 repeats is very likely to develop problems such as mental retardation, hyperactivity, heart valve defects, and other problems; women are often affected in a milder way. (They have a second copy of the gene on their second X chromosome, which may be unaffected.) Which symptoms develop and their severity differ from case to case. Since the problem lies on the X chromosome, it is passed from mother to child, and it frequently becomes more severe from generation to generation. Once the gene has become unstable in a mother, she may produce egg cells with far more repeats.

Repeats of small DNA sequences within other genes have now been linked to other developmental problems and life-threatening conditions. Huntington's disease, the symptoms of which include a progressive loss of muscle control and often mental abilities, is caused by extra repeats in the huntingtin gene. It normally holds between nine and about 34 repeats of the sequence "CAG." Some people inherit many more, and if the number is too high they develop the severe neurodegenerative condition. In itself the disease is not fatal, but it usually leads to complications that result in death. Another case is myotonic muscular dystrophy, a muscle-wasting disease that is accompanied by problems in the relaxation of muscles and heart defects.

Fragile X syndrome is a dramatic example of how mutations in a single gene can disrupt the development of the brain and disturb cognitive functions. Researchers are also interested in the links between genes and other, usually milder defects in mental processes. It is becoming clearer, for example, that many cases of autism and several related problems, called autism spectrum disorders (ASD), have a genetic component. Autism, which usually appears in the early stages of childhood development, is a condition in which a person has problems interacting and communicating with others, empathizing with them, and developing friendships and normal social interactions; it is also often accompanied by physical awkwardness and repetitive behavior. Physicians now regard it as the most severe form of a spectrum of behavior that, in some people, manifests itself much more mildly.

Studies of twins and families with multiple cases of ASD seem to reveal that several genes may be involved. This is one thing that makes them difficult to find (for the reasons described in the section "Multifactorial Disease" in chapter 4). Another is the lack of accurate diagnoses—the symptoms are much milder in some than in others, and they vary so widely from person to person that many cases probably go unnoticed, or are wrongly classified. An accurate diagnosis is essential to linking diseases or other types of developmental problems to genes. There is also an important social issue. Some people who are affected—and their families—do not regard ASDs as disorders. They point out that many of history's great artists and thinkers may have had some form of ASD, and that special features of their brains may be responsible for both great talents and difficulties in coping with other aspects of life.

But severe manifestations of the symptoms—particularly in autism—usually pose a great challenge to people with the disorder and their families. There is currently no treatment, although behavioral therapies can ease the struggle that many ASD sufferers have in coping with social situations and in learning to lead independent lives. Creating effective therapies will probably depend on finding genes that are involved and understanding how they influence the functions of cells and the development of the brain.

In 2008 Brian O'Roak and Matthew State of Yale University summarized the state of the art in ASD genetics in an article in the first issue of a new online journal called *Autism Research.* They pointed out that large-scale studies of families have shown that children with an autistic sibling have a 2 to 8 percent risk of being autistic themselves, which is between 20 and 80 times higher than the risk in the general population. These numbers are about the same for other forms of ASD. Twin studies also support the idea of a genetic component to the diseases. However, the authors report that "over the course of the past decade, large-scale genetic studies have effectively ruled out the possibility that a single gene of large affect will be found to account for a substantial portion of ASD." In some families rare, specific mutations seem to trigger the disease, but different families have different ones. Thus there seem to be many paths that lead to similar symptoms. The more distantly related a family member is to an affected person, the lower his risk. The authors conclude that "there are several lines of evidence suggesting that a significant proportion of ASD may be the result of a conspiracy of common risk alleles within an individual, typically in concert with environmental influences, leading to the phenotype."

Some of the genes that have been identified are the following:

Neuroligin 4 X-linked (NLGN4X): a gene on the X chromosome. The protein it encodes is found on the surface of neurons and is likely involved in the formation and remodeling of synapses, which are crucial to learning, memory, and overall cognitive abilities.

MECP2: also on the X chromosome. Some mutations cause Rett syndrome, whose symptoms include the development of a small head and hands and is often accompanied by cognitive defects and impaired social skills. The link between the gene and autism is not surprising because of this overlap of some of their symptoms.

Contactin-Associated Protein-Like 2 (CNTNAP2): one of the largest human genes, located on chromosome 7. CNTNAP2 encodes a receptor protein involved in cell adhesion and is thought to help in the differentiation of neurons. Defects have been linked to childhood epilepsy and mental retardation.

FMR1: the fragile X syndrome gene. ASD is diagnosed in 30 percent of males with FXS, and 7 to 8 percent of people with ASD have mutations in genes. Here again, it is not surprising to find a link to ASD because of the molecule's role in the transmission of information between nerve cells.

MET: mesenchymal-epithelial transition factor, located on chromosome 7. The MET protein is a receptor involved in the development of a wide range of tissues in the body, in processes such as wound healing, and in cancer. A mutation that reduces MET's activity is common in autistic children.

This list is only a beginning. At the time of writing there is a large-scale effort underway to uncover new genes related to ASD. In 2007 three new "hot spots" in the human genome were identified by researchers of the Autism Genome Project Consortium—consisting of laboratories from Johns Hopkins University, the University of California at San Francisco, Duke University in Durham, North Carolina, the University of Washington in Seattle, and many other institutions. They are collaborating on a massive project involving nearly 1,600 trios of parents and autistic children, looking for SNPs that are linked to the symptoms of ASD. One discovery is a site on chromosome 5 that may help protect people from the disease. The DNA sequence lies near a gene called semaphorin; one of its functions is to help connect nerves to each other during embryonic development. On the whole, only about 4 percent of the population has the particular sequence uncovered by the project. But autistic children have it at an even lower rate. The researchers hypothesize that if the children had inherited this "healthy" version of the sequence, they might not have developed the disorder.

Logically, most of the molecules that have been linked to ASD are involved in the "hard wiring" of the brain—establishing connections between neurons—or communication between nerve cells. In some individuals, mutations in these molecules arise spontaneously; once they have entered the genome, they can be passed along through heredity. In many cases it is likely that additional genetic or environmental factors are necessary for these gene defects to cause ASD. The Autism Genome data

Kim Peek and the Savants

Kim Peek (1951–) is one of a small group of people with "savant syndrome," a rare condition in which unusual brain development and disabilities in some areas are accompanied by exceptional talents in others. Kim came to the attention of the world when American director Barry Levinson made the film *Rain Man*. In the movie Dustin Hoffman delivered an Oscar-winning performance as a savant named Raymond Babbit, a middle-aged man with very limited social skills, who has an autistic-like condition and has been institutionalized for most of his life. Upon the death of his father, Raymond is reacquainted with his younger brother Charlie (played by Tom Cruise) and plunged into the real world. The brothers embark on a road trip in which Charlie becomes aware of Raymond's photographic memory and uncanny mathematical skills. In 1984 screenwriter Barry Morrow met Kim and wrote the film's script (which also won an Oscar) largely with him in mind.

Kim's special abilities include an almost perfect memory for the contents of books he has read and pieces of music he has heard. He is also adept at calendar calculations—given a particular date, he can almost instantly say the day of the week on which it falls, whether it is in the remote past or distant future. His astounding memory also permits him to remember headlines from newspapers for each day stretching far into the past.

With the appearance of the film, Kim's story became known throughout the world, and his condition and that of other savants have captured the imagination of the general public and become a topic of greater interest in the medical community. Kim has become a much sought-after speaker, traveling around the country with his father Fran Peek and giving talks to enthusiastic audiences.

His many encounters with strangers have helped him develop new social skills, as well as helping expand others' appreciation and understanding of people who are born "different."

Doctors have established that Kim was born with a condition called macrocephaly, in which the brain and head grow to an unusually large size—in his case, 30 percent larger than average at the time of birth. Doctors also found that he lacks a brain structure called the corpus callosum. This tissue usually consists of more than 200 million axons that connect neurons in one hemisphere of the brain to the other; it is important in coordinating body movement and in cognitive functions. Researchers believe that instead of crossing between the hemispheres, the axons of Kim's brain probably connect to cells in different parts of the brain, and this may partly explain his ability to remember vast amounts of information. In 2008 researchers at the University of Utah in Salt Lake City, where Kim lives, established that he probably has FG syndrome, a rare genetic condition caused by a mutation in a gene called MED12, on the X chromosome.

Psychiatrist Darold Treffer of the University of Wisconsin Medical School has been studying savant syndrome for nearly four decades and is likely the world's foremost expert on this condition. Each savant is unique, in terms of the genetic cause of the condition, the way the brain develops as a result, and the types of skills that arise. Some are magnificent painters, sculptors, or poets. Others are gifted in mathematics or languages. Many have combinations of skills. Daniel Tammet, a British man born in 1979, sees numbers as colors and shapes, learned Icelandic in a week, and once recited the first 22,514 digits of pi from memory over a five-hour period.

(continues)

(continued)

Although many people with savant syndrome are now known, they are very rare. Treffer estimates that about 10 percent of people with ASD have savant skills, but the National Autistic Society of the United Kingdom puts the number much lower, at 1 percent or less. Most are male, and about half are autistic. The rest have other types of brain damage. Another type of individual, known as a "prodigious savant," has similar skills but without any obvious cognitive disability. This condition, too, is extremely rare.

shows that some people may carry the mutations but escape symptoms because they are protected by other factors.

The same challenges face those who hope to explain other aspects of complex brain function and behavior, whether healthy or unhealthy. For decades researchers have wondered to what extent abilities such as exceptional musical or mathematical skills—as well as violence or other antisocial tendencies—depend on genes rather than environmental influences. While everything a human does depends on both, it may still be interesting and useful to tie behavior to specific genes, particularly when it comes to diagnosing and treating mental disorders.

A single human cell can develop in hundreds of different ways, depending on its environment. An organ like the human brain—which contains trillions of cells, linked in slightly different ways in each individual—is so complex that researchers may never be able to completely tell which parts of a behavior are due to genes and which are due to the environment. The search for the causes of ASD shows how difficult it will be to answer these questions.

HUMAN ADAPTATION AND COMPETITION BETWEEN SOCIETIES

Recent centuries have seen the rise of huge differences between the developed world, with its comforts and high-tech lifestyle, and developing countries, where poverty, famine, and infectious diseases are still daily facts of life. These differences had become extreme by the mid-19th century and had touched off debates about the differences between races, the origins of government, the future of human progress, and a wide range of other social themes. Western philosophers proposed answers based on their religious beliefs or moral systems—for example, the idea that industrial societies had the "right" religion and were favored by God, or that human societies were climbing a ladder to moral perfection, with some groups or races farther along than others. With the arrival of the theory of evolution, people began to wonder if biology might explain some of the differences between human societies. Jared Diamond (1937–), who is professor of geography and physiology at the University of California at Los Angeles, has put these questions into a modern context by looking at the contribution that geography, genetics, and other factors have made to recent human history. In doing so, he tries to establish a link between human genes and the competition between cultures.

Even before the theory of evolution, there had been attempts to find a biological basis for inequalities in societies. In an article published in 1857, the influential British philosopher Herbert Spencer (1820–1903) presented a view of the universe as "evolving" from a simple, uniform state to one of more complexity and diversity. This included living creatures, human beings, cultures, and behavior, and Spencer believed they were all undergoing a process of continual improvement. In "barbarous tribes," he claimed, every person had similar functions; as society advanced, certain strong figures became leaders. This led to a "differentiation between the governing and the governed," followed by more and more social distinctions. Spencer saw the universe and everything in it as something like a human body, which begins as something simple and undifferentiated but then increases in complexity and develops highly specialized tissues. And just as human cells are programmed to become an adult human being, he believed that society is developing toward an ultimate form.

A year after the appearance of Spencer's article, Charles Darwin and Alfred Russel Wallace introduced the idea that natural selection had produced new species, including humans. Spencer quickly saw that this "natural law," for which he coined the phrase "survival of the fittest," might provide support for some of his opinions. He began using the new concepts to talk about society. Like many of his contemporaries, he confused biological "success" (which simply meant that hereditary features prompted some organisms to reproduce more successfully than others) with marks of success within society—wealth, strength, and technological progress. Spencer also believed that as organisms and society evolved, they continually improved themselves.

Despite these fundamental differences, many thinkers wove biological ideas into a muddled concept of "Social Darwinism," which influenced how scientists tried to understand technological progress, social change, and differences between cultures. The confusion lasted for many decades; only in modern times have sociology and political science become relatively free of moral judgments and Western bias.

Jared Diamond's new approach to society and culture is much more closely based on Darwin's ideas and modern biological discoveries. His 1998 book *Guns, Germs, and Steel* aims to explain how Euroasian countries came to dominate the globe because of superior weapons, other types of technology, and the deadliness of the diseases they brought along. To find answers he reaches back far into modern human history—an alternative title for the book, he writes, would be a history of "everyone for the last 13,000 years."

One factor that has continually played a role in historical conflicts is the fact that Euroasian travelers, settlers, and armies brought along diseases that decimated populations in the new lands. Smallpox, carried by the first Spaniards to reach the New World, killed a huge proportion of the population, including two Mexican emperors. This destabilized the Aztec political system and weakened its army, allowing the conquistador Hernán Cortés to achieve military victories in the face of overwhelming odds. In North America, smallpox arrived with the first European settlers and likewise devastated Native American populations. In modern times, influenza and other diseases have had a similar effect on remote tribes in the Amazon.

This has not been a one-way situation; the travelers also were afflicted by diseases of the lands they visited. When settlers returned to Europe, they carried along syphilis and other epidemics that had originated elsewhere. But generally the diseases spread by the Europeans were far more deadly and gave them an advantage in their conquest of other lands. Diamond's hypothesis to explain this is that geography and other features of the environment have combined to help give Euroasians stronger immune systems than people living in other regions of the world. To understand why, it is necessary to take a deep look at the factors that influence disease organisms and their hosts.

Many authors have called this relationship an "evolutionary arms race," referring to the way organisms coevolve as the result of natural selection. Viruses and many other infectious agents need a host to survive. Usually this is one species, or a small group of closely related ones. If a disease is so terrible that it completely wipes out the host species, the pathogen may die as well. This may have happened many times in history—it may be the cause of some of the extinctions seen in the fossil record. But if a disease has survived a long time, the reason is likely to be that it and its host have adapted to each other. It still may make a plant or animal sick, but its own survival may well depend on the host living long enough to have new offspring for it to infect. Nor will it survive if the host's immune system is strong enough to defeat it before it can jump to another member of the species. This, too, has probably happened over the course of history; it is hard to know, because the pathogens are no longer around to study. So today's diseases—at least the old ones—have usually achieved a sort of balance with their hosts. Of course, they continue to undergo mutations that may produce much more deadly versions.

Viruses or other pathogens that cross species barriers have not gone through this long process of adaptation in the new host. On one hand this means they may be much more deadly than an animal's normal parasites. On the other, they are likely to have a harder time surviving in its body. Because of the normal variation among members of a species—thanks to different forms of genes inherited from their parents, or new mutations they have experienced—some plants or animals will be better than others at warding off the disease. If they pass "resistance genes" along to their offspring, the species as a whole will eventually gain a degree of protection. Again, if this is too perfect, the pathogen may become extinct.

Diamond points out that Europe and Asia have an open landscape, with few significant geographical barriers to inhibit migrations, travel, and trade along the East-West axis. This made it easy for early agricultural communities to spread because they could remain in the same latitude where the same crops could be grown. Migrations and trade continually brought Euroasian groups in contact with each other, which also gave them exposure to each others' diseases. These continual challenges led to the development of a stronger immune system that could cope with a wide range of threats.

The geography of the Americas, which was settled much later, did not offer the same advantages. The long axis of these two continents stretches north and south; as a result, as populations moved, they needed to learn to raise different plants and animals because of significant changes in climate. Natural barriers such as mountains, deserts, and rain forests inhibited the sorts of massive migrations and continual travel between regions. Groups that became more isolated adapted to a small set of diseases and faced fewer challenges. There were fewer animals suitable for domestication. Thus, when the natives came into contact with Spaniards or other visitors from the Old World, their immune systems were at a disadvantage.

The plants and animals available for cultivation have also had a major effect on how societies have developed in different regions of the world. For example, wheat—which probably arose in the Middle East as a hybrid of wild grasses—was both nutritious and easy to sow. The major crop of the Americas—corn—has fewer nutrients and is harder to plant, which limits the size of the community that it can support. Domesticated animals have played an even greater role. While many animals of Asia and Europe—cattle, goats, sheep, pigs, and chickens—were relatively easy to domesticate, those of sub-Saharan Africa were larger, wilder, and more dangerous. Close contact with domesticated species was a major source of diseases that threatened European and Asian populations but simultaneously improved their bodies' defenses.

Diamond weaves these facts into a convincing argument that these ancient differences in geography, climate, and resources had a strong effect on human biology and society, and is the main explanation for recent imbalances in the distribution of wealth around the world. This is a much different picture than the one painted by Herbert Spencer and other early "Social Darwinists"; Diamond's version is free from judgments about the goal of society or the moral superiority of some cultures over others.

Conclusion

A Look into the Future: The Genetic Engineering of Human Beings?

The 1980s saw the beginnings of a revolution in biotechnology that has given scientists powerful new tools to understand how genes influence organisms' lives. Genetic engineering allows scientists to alter the DNA of a cell, plant, or animal. By the end of the 20th century, it had become routine to make deliberate, targeted changes in plants, animals, and human cell lines. In principle, the same technology can be applied to humans. Currently, the idea of directly engineering the human genome is considered unethical by nearly everyone. But will that also be the case at the end of the 21st century?

DNA changes constantly through natural mutations, and in the early days of genetic science, researchers had to wait for them to happen naturally in the laboratory organisms they wanted to study. That began to change in the late 1920s, when Hermann Muller showed that radiation could increase the rate at which mutations occurred. This was soon followed by the discovery of other *mutagens* (such as chemicals) that could accomplish the same thing. As useful as these techniques were, they all had a drawback: The changes they caused in genes were random and unpredictable. If they affected an important gene, it would likely have dramatic effects on the organism's body. Then the researcher had to work backward, trying to guess what process had been disturbed and use linkage studies to try to pin down the gene's position on a chromosome.

In the early 1980s two laboratories in California launched the era of genetic engineering when they found a way to make precise changes in specific DNA sequences and even to transplant genes between species. While these techniques have not been used on egg or sperm cells to change the human genome, they have been used on human cells in the test tube and are now routinely used to investigate laboratory animals. Most of what is known about the functions of human genes has come from such experiments with close evolutionary relatives like the mouse, as well as more distant cousins like the fruit fly.

A few key discoveries set the stage for genetic engineering. In the late 1950s a Swiss scientist named Werner Arber (1929–   ) was investigating how bacteria become resistant to attacks by viruses called phages. Salvador Luria (1912–91), working at the Massachusetts Institute of Technology, had discovered that bacteria had proteins called *restriction enzymes* that helped protect them from the virus. Arber and Hamilton Smith (1931–   ), of Johns Hopkins University, showed that the proteins formed part of a bacterial defense system. In order to reproduce, viruses have to get the cell they have invaded to copy their DNA. Restriction enzymes in the bacteria recognized that the DNA was foreign and chopped it into small pieces.

Genetic engineering requires a pair of "scissors" (to remove a gene from one place) and a sort of glue (to paste it in somewhere else). Restriction enzymes provided the scissors. Bacteria contained another type of molecule, called a *ligase,* that could bind the broken ends together again and mend the cuts. Such enzymes are important because DNA sometimes breaks by mistake. Ligases can scout the molecule and make repairs by matching up the broken ends. Sometimes they glue the wrong breaks together, leading to the loss of a gene or a scrambling of genetic information. This provided the second tool that genetic engineers needed—a glue to paste new DNA sequences into genomes.

In 1972 Janet Mertz and Ron Davis, of Stanford University, combined restriction enzymes and ligases to create the technique now known as DNA recombination. A year later, Herbert Boyer of the University of California at San Francisco and two colleagues at Stanford University, Stanley Cohen and Annie Chang, put the method to work to move a gene from one species to another. They combined genetic material from a virus and a bacteria and inserted it into another bacteria. This artificial gene was taken up by the cell and used to create a foreign protein. Ironically, to transplant genes across species, the scientists were making use of molecules that had almost certainly evolved to prevent foreign DNA from invading cells. These accomplishments ushered in the age of genetic engineering, which is now a fundamental tool in practically every type of biological research and has many applications in the production of food, drugs, and many other fields.

Modern techniques allow scientists to remove a gene, substitute another one for it, or add a new molecule to an organism's genome. By watching what happens as the animal develops or how it responds to a disease, researchers can learn the roles that a specific molecule plays in its cells and tissues. Once they learn these things, scientists can begin to alter the molecule's functions in specific ways. If the molecule has a medical use, the methods permit scientists to use other species to produce it.

For many years bacteria and animals have been routinely used to make insulin, a human protein used to treat diabetes. Insulin cannot be obtained in large amounts from humans. Doctors used to administer molecules extracted from pigs or cows, but their bodies produce a slightly different form of the molecule, which sometimes causes rejection by the human immune system or long-term health problems. Changing the recipe of the animals' insulin genes leads them to make a more human version that can be safely used.

Genetic engineering has many other practical uses. A rising percentage of the corn, tomatoes, soybeans, rice, and dozens of other crops produced across the world have been manipulated through genetic engineering. As well as attempting to improve the size, taste, shelf life, or nutritional value of crops, scientists have transplanted genes that help protect plants from insects, fungi, and other parasites. These changes might help farmers ward off pests without the dangerous side effects of pesticides. On the other hand, the members of growing ecological and environmental movements protest that genetic engineering might upset delicate balances in nature.

The debate continues and it is an important one. Although plants with modified genes undergo rigorous tests before they leave the laboratory, their impact on the environment cannot be completely foreseen. So tests may never be able to push aside all the concerns that people have about the consequences of altering an organism's genetic material and releasing it into nature.

But fears should be kept in perspective. Normal farming techniques also alter the genes of plants and animals, sometimes in far more dramatic ways, but the changes are almost never tested for possible harmful effects. Rarely is there an investigation of which genes have been lost, added, or changed through these practices—the technology to do so has only existed for a few years, and the cost would be enormous. In the wild, new genes appear in plants all the time because of mutations, and they may even be naturally transferred between species by viruses or other parasites. When the pollen of one plant species fertilizes the egg of another, the result may be a monstrous collection in which hundreds or thousands of new genes collide for the first time and interact in unpredictable ways.

These "natural cases of genetic engineering" have exactly the same potential to cause harm as organisms produced in the laboratory—in fact, they may be much more dangerous because there have been no controls on their production and release. This means that many of the fears about genetically modified organisms are not based on experience or available facts. Instead, they probably stem from religious or philosophical concerns about man's relationship to nature, and memories of the detrimental effects that other types of technology have had on the environment.

While the idea of creating genetically modified humans is considered unethical, many feel that genetic testing and experimental medical therapies are blurring the line. Chapter 4 describes gene therapies and other techniques whose aims are to correct hereditary defects by delivering healthy versions of genes to cells. If these attempts are successful, they will alter the genome of some of a patient's cells. These will be somatic changes, however, that do not enter the genetic material of egg and sperm cells, which means they will not be inherited. While genetic engineering is involved, the treatments affect only specific patients, and not their offspring, so most people regard such interventions as "normal medicine."

The next step—directly intervening in human evolution by manipulating the genome—may never be permitted by society. The idea is profoundly disturbing to those who remember the sterilization programs of "unfit" people in the United States and many other countries, and the Holocaust—terribly misguided attempts to "improve" the human race.

But if a family were offered the chance to correct a dangerous gene—for example, the defect that causes Huntington's disease—would the desire to prevent future generations from suffering override their ethical concerns? Most researchers believe that this will be possible in the not-too-distant future, certainly within the lifetime of today's young students. Such techniques will be part of the new world that science is creating, a world in which society will confront new ethical problems and difficult choices. The only way to be prepared is to understand the issues ahead of time, and to consider them clearly and deeply with a profound respect for human life.

Chronology

1651   William Harvey claims that all animals arise from eggs.

1677   Anton van Leeuwenhook discovers sperm.

1751   Pierre-Louis Maupertius studies polydactyly, the inheritance of extra fingers in humans.

Joseph Adams recognizes the negative hereditary effects of inbreeding.

1802   Jean-Baptiste Lamarck publishes *Research on the Organization of Living Bodies,* in which he claims that species change through environmental influences and through their own activities and behavior. His philosophy of species change is based on the notion that animals seek to become more perfect and better adapted to their surroundings.

1817   Georges Cuvier publishes *The Animal Kingdom,* in which he argues that all the characteristics of a species are attuned to fit its lifestyle. His studies of fossils show that they were living creatures that have become extinct.

1820s   Étienne Geoffroy St. Hilaire's studies of animal anatomy point out unusual relationships between species and features that seem to have lost their functions over time.

1824   Joseph Lister builds a new type of microscope that removes distortion and greatly increases resolution.

1827   Karl Ernst von Baer is first to discover an egg cell in a mammal (a dog).

1830   Giovanni Amici discovers egg cells in plants.

1838   Matthias Schleiden states that plants are made of cells.

1840   Theodor Schwann states that all animal tissues are made of cells.

1855   Rudolph Virchow creates the cell doctrine: "All cells arise from pre-existing cells."

1856   Gregor Mendel begins experiments on heredity in pea plants.

1857   Joseph von Gerlach discovers a new way of staining cells that reveals their internal structures.

1858   The theory of evolution is made public at a meeting of the Linnean Society in London with the reading of papers by Charles Darwin and Alfred Russel Wallace.

Rudolf Virchow states the principle of *Omnis cellula e cellula*: every cell derives from another cell—including cancer cells.

1859	Charles Darwin publishes *On the Origin of Species.* The complete first print sells out on the first day.
1865	Gregor Mendel presents his paper "Experiments in Plant Hybridization" in meetings of the Society for the Study of Natural Sciences in Brnø, Moravia. The paper outlines the basic principles of the modern science of genetics. It is published the next year but receives little attention.
1868	Fredrich Miescher isolates DNA from the nuclei of cells; he calls it "nuclein."
1871	Francis Galton carries out experiments in rabbits that disprove Darwin's hypothesis of how heredity functions.
1876	Oscar Hertwig observes the fusion of sperm and egg nuclei during fertilization.
1879	Walther Flemming observes the behavior of chromosomes during cell division.
1885	August Weismann states that organisms separate reproductive cells from the rest of their bodies, which helps explain why Lamarck's concept of evolution and inheritance is wrong. He tries and fails to observe Lamarckian inheritance in the laboratory by cutting off the tails of mice for many generations.
1894	Emil Fischer describes the fact that specific enzymes recognize and change each other using the metaphor of locks and keys.

1900	Hugo de Vries, Carl Correns, and Erich Tschermak von Seysenegg independently publish papers that confirm Mendel's principles of heredity in a wide range of plants.
	Archibald Garrod first identifies a disease that is inherited according to Mendelian laws, which means that it is caused by a defective gene.
	Theodor Boveri demonstrates that different chromosomes are responsible for different hereditary characteristics.
1901	Karl Landsteiner identifies the ABO blood groups.
1902	William Bateson popularizes Mendel's work in a book called *Mendel's Principles of Heredity: A Defense.*
	Archibald Garrod publishes his finding that a form of arthritis follows a Mendelian pattern of inheritance.
1903	Walter Sutton connects chromosome pairs to hereditary behavior, demonstrating that genes are located on chromosomes.
1905	Nettie Stevens and Edmund Wilson independently discover the role of the X and Y chromosomes in determining the sex of animal species.
1906	William Bateson discovers that some characteristics of plants depend on the activity of two genes.

1908	Archibald Garrod shows that humans with an inherited disease are lacking an enzyme (a protein), demonstrating that there is a connection between genes and proteins.
1909	William Bateson coins the term "genetics."
1910	Eugenics Record Office opened at Cold Spring Harbor, New York.
	Thomas Hunt Morgan discovers the first mutations in fruit flies, *Drosophila melanogaster,* bred in the laboratory. This leads to the discovery of hundreds of new genes over the next decades.
1911	Morgan discovers some traits that are passed along in a sex-dependent manner and proposes that this happens because the genes are located on sex chromosomes. He proposes the general hypothesis that traits likely to be inherited together are located on the same chromosome.
1913	Alfred Sturtevant constructs the first genetic linkage map, allowing researchers to pinpoint the physical locations of genes on chromosomes.
1920	Hans Spemann and Hilde Proescholdt Mangold begin a series of experiments in which they transplant embryonic tissue from one species to another. The scientists show that particular groups of cells they called "organizers" sent instructions to neighboring cells that changed their developmental fates.

1921	Erwin Baur, Eugen Fischer, and Fritz Lenz publishes *Menschliche Erblichkeitslehre und Rassenhygiene,* which becomes the bible of human genetics instruction in Europe and the United States, as well as the handbook to the Nazi eugenics program.
1922	Ronald A. Fisher uses mathematics to show that Mendelian inheritance and evolution are compatible.
1924	Theophilus Painter establishes the human chromosome number as 48.
1927	Hermann Muller shows that radiation causes mutations in genes that can be passed down through heredity.
1928	Fredrick Griffith discovers that genetic information can be transferred from one bacterium to another, hinting that hereditary information is contained in DNA.
1931	Barbara McClintock shows that as chromosome pairs line up beside each other during the copying of DNA, fragments can break off one chromosome and be inserted into the other in a process called recombination.
	Archibald Garrold proposes that diseases can be caused by a person's unique chemistry—in other words, genetic diseases may be linked to defects in enzymes.

1933	Theophilus Painter discovers that staining giant salivary chromosomes in fruit flies reveal regular striped bands.
1934	Calvin Bridges shows that chromosome bands can be used to pinpoint the exact locations of genes.
1935	Nikolai Timofeeff-Ressovsky, K. Zimmer, and Max Delbrück publish a groundbreaking work on the structure of genes that proposes that mutations alter the chemistry and structure of molecules.
1937	George Beadle and Boris Ephrussi show that genes work together in a specific order to produce some features of fruit flies.
1940	George Beadle and Edward Tatum prove that a mutation in a mold destroys an enzyme and that this characteristic is inherited in a Mendelian way, leading to their hypothesis that one gene is related to one enzyme (protein), formally proposed in 1946.
1943	Max Delbrück and Salvador Luria demonstrate evolution in the laboratory by showing that bacteria evolve defenses to viruses through mutations that are acted on by natural selection.
1944	Oswald Avery, Colin MacLeod, and Maclyn McCarty show that genes are made of DNA.

Erwin Schrödinger publishes *What Is Life?* |

1948   The American Society for Human Genetics is founded.

1950   Barbara McClintock publishes evidence that genes can move to different positions as chromosomes are copied.

Erwin Chargaff discovers that the proportions of A and T bases in an organism's DNA are identical, as are the proportion of Gs to Cs.

1951   Rosalind Franklin uses X-ray diffraction to obtain images of DNA; the patterns reveal important clues to the building plan of the molecule.

1953   James Watson and Francis Crick publish the double-helix model of DNA, which explains both how the molecule can be copied and how mutations might arise.

In the same issue of the journal *Nature,* Rosalind Franklin and Maurice Wilkins publish X-ray studies that support the Watson-Crick model. This launches the field of molecular biology, which shows, over the next 20 years, how the information in genes is used to build organisms.

1958   Francis Crick describes the "central dogma" of molecular biology: DNA creates RNA creates proteins. He challenges the scientific community to figure out the molecules and mechanisms by which this happens.

1959   The first chromosomal disease is identified: Jerome Lejeune discovers that Down syndrome is caused by the inheritance of an extra chromosome.

Marshall Nirenberg, Marianne Grunberg-Manago, and Severo Ochao show that the cell reads DNA in three-letter "words" to translate the alphabet of DNA into the 20-letter alphabet of proteins.

1961   Sidney Brenner, François Jacob, and Matthew Meselson discover that messenger RNA is the template molecule that carries information from genes into protein form. Crick and Brenner suggest that proteins are made by reading three-letter codons in RNA sequences, which represent three-letter codes in DNA. M. W. Nirenberg and J. H. Matthaei use artificial RNAs to create proteins with specific spellings, helping them learn the complete codon spellings of amino acids.

1966   Marshall Nirenberg and H. Gobind Khorana work out the complete genetic code—the DNA recipe for every amino acid.

1970   Hamilton Smith and Kent Wilcox isolate the first restriction enzyme, a molecule that cuts DNA at a specific sequence—this will become an essential tool in genetic engineering.

1972   Janet Mertz and Ron Davis use restriction enzymes and DNA-mending molecules called ligases to carry out the first recombination: the creation of an artificial DNA molecule.

Paul Berg creates a new gene in bacteria using genetic engineering.

1973   Stanley Cohen, Annie Chang, Robert Helling, and Herbert Boyer create the first transgenic organism by putting an artificial chromosome into bacteria.

1975   Edward Southern creates Southern blotting, a method to detect a specific DNA sequence in a person's DNA; the method will become crucial to genetic testing and biology in general. Cesar Milsein, Georges Kohler, and Niels Kai Jerne develop a method to make monoclonal antibodies.

1977   Walter Gilbert and Allan Maxam develop a method to determine the sequence of a DNA molecule; Fredrick Sanger and colleagues independently develop another very rapid method for doing so, launching the age of high-throughput DNA sequencing.

Frederick Sanger finishes the first genome, the complete nucleotide sequence of a bacteriophage.

Phillip Sharp and colleagues discover introns, information in the middle of genes that do not contain codes for proteins and must be removed before an RNA can be used to create a protein.

Genentech, the first biotech firm, is founded based on plans to use genetic engineering to make drugs.

1978   Recombinant DNA technology is used to create the first human hormone.

1980   Christiane Nüsslein-Volhard and Eric Wieschaus discover the first patterning genes that influence the development of the fruit fly embryo, bringing together the fields of developmental biology and genetics.

1981   Three laboratories independently discover oncogenes, proteins that lead to cancer if they undergo mutations.

1982   Insulin becomes the first genetically engineered drug.

1983   Walter Gehring's laboratory in Basel, Switzerland, and Matthew Scott and Amy Weiner at Indiana University in Bloomington, independently discover HOX genes: master patterning molecules for the creation of the head-to-tail axis in animals as diverse as flies and humans.

1985   Kary B. Mullis publishes a paper describing the polymerase chain reaction, a method that rapidly and easily copies DNA molecules.

1986   The first outbreak of BSE (mad cow disease) among cattle in Great Britain.

1987   First human genetic map published.

1988   The Human Genome Project is launched by the U.S. Department of Energy and the National Institutes of Health, with the aim of determining the complete sequence of human DNA.

1989   Alec Jeffreys discovers regions of DNA that undergo high numbers of mutations. He develops a method of "DNA fingerprinting" that can match DNA samples to the person from which they came and can also be used in establishing paternity and other types of family relationships.

The Human Genome Organization (HUGO) is founded.

1990   W. French Anderson carries out the first human gene replacement therapy to treat an immune system disease in four-year-old Ashanti DeSilva.

1993   The company Monsanto develops and begins to market a genetically engineered strain of tomatoes called FlavrSavr.

1993   The Huntington's disease gene is discovered.

1994   Mary-Claire King discovers BRCA1, a gene that contributes to susceptibility to breast cancer.

1995   The first confirmed death from Creutzfeldt-Jakob disease, the human form of BSE, is reported in Great Britain.

1996	Researchers complete the first genome of a eukaryote, baker's yeast. The completion of the genome of *Methanococcus jannaschii,* an archaeal cell, confirms that archaea are a third branch of life, separate from bacteria and eukaryotes.
	Gene therapy trials to use the adenovirus as a vector for healthy genes are approved in the United States.
1997	Ian Wilmut's laboratory at the Roslin Institute produces Dolly the sheep, the first cloned mammal.
1998	Scientists obtain the first complete genome sequence of an animal, the worm *Caenorhabditis elegans.*
1999	Jesse Gelsinger dies in a gene therapy trial, bringing a temporary halt to all viral gene therapy trials in the United States.
2000	The genome of the fruit fly, *Drosophila melanogaster,* is completed.
	Scientists complete a "working draft" of the human genome. The complete genome is published in 2003.
2002	The mouse genome is completed.
2004	Scientists in Seoul, South Korea, announce the first successful cloning of a human being, a claim that is quickly proven to be false.

2008   Samuel Wood, of the California company Stemagen, successfully uses his own skin cells to produce clones, which survive five days.

Glossary

adaptive immune system a network of cells and body structures in vertebrates that is capable of recognizing, fighting, and remembering new threats from parasites or other infectious agents

adenine one of the nucleic acids that make up DNA and RNA molecules

albinism an inherited condition in which a person or animal lacks a pigment called melanin, usually in the eyes, skin, and hair

allele one variant of a single gene within a given species

Alu sequences short stretches of DNA with a common motif that make up a huge proportion of human and primate genomes, thought to have originated as "jumping genes"

Alzheimer's disease a neurodegenerative condition that usually strikes in old age and results in the gradual loss of cognitive skills and control of the body

amino acid the fundamental chemical subunit of a protein. There are 20 types, built around an identical core of carbon, hydrogen, oxygen, and nitrogen atoms, made unique by a side chain of other atoms.

antibody a molecule on B cells that plays a key role in adaptive immunity. Antibodies are produced through random rearrangements of genes, creating a huge range of structures that can recognize foreign molecules or substances. They mark invading viruses or microbes for destruction by immune system cells.

antigen a molecule recognized by an antibody

archaea   single-celled organisms that are thought to be the oldest types of cells on Earth, ancestors of bacteria and eukaryotes. Many are extremophiles, living at high temperatures or in other extreme environments.

autism   a developmental brain disorder that is accompanied by a variety of symptoms including problems with communication and social interactions, as well as repetitive behavior

axon   a long extension growing from a neuron which stimulates neighboring cells

B cell receptor   an antibody on the surface of a B cell (a type of white blood cell) that triggers a response from the immune system if it docks onto a foreign molecule

base   a nucleotide, a subunit of a DNA molecule which binds to a partner nucleotide on a complementary strand of DNA

centrosome   a small cellular structure also known as a "microtubule organizing center" that acts as a pole of the mitotic spindle during most types of cell division

chromosome   a large, compressed cluster of DNA and many other molecules found in the cell nucleus of eukaryotes

codon   a three-letter sequence in RNAs that is used to translate the four-letter alphabet of DNA and RNA into the 20-letter amino acid code

constant region   subregions of antibody molecules that are identical. They combine with variable regions to create billions of types of antibodies with unique structures.

cytoplasm   the major compartment of the cell, which surrounds the nucleus and contains organelles

cytosine   one of the nucleic acids that make up DNA and RNA molecules

cytoskeleton   a system of protein tubes and fibers that give the cell shape and structure and play a key role in processes like cell division and migration

dendrite a branchlike network of extensions that grow from neurons and are stimulated by chemical signals from neighboring cells

DNA (deoxyribonucleic acid) a molecule made of nucleic acids that forms a double helix in cells, holds a species' genetic information, and encodes RNAs and proteins

dominant in a situation where an organism has different alleles of the same gene, the dominant allele is the one that influences what happens in cells and the body, and a recessive one does not. Having a single copy of one dominant allele has the same effect as having two copies.

ectoderm the outermost of the three main layers of tissue that arise during the process of gastrulation in early embryonic development, or tissues that arise from this layer

embryonic stem cells cells that arise at the very beginning of an organism's development and have the capacity to differentiate into all other types of cells in its body

endoderm the innermost of the three main layers of tissue that arise during the process of gastrulation in early embryonic development, or tissues that arise from this layer

eubacteria another term for bacteria

eugenics an attempt to change the genetic makeup of the human species by selecting particular people to reproduce, preventing others from reproducing, or by selecting certain fetuses or babies over others for survival

eukaryote member of one of the three major domains of life (the other members are archaea and bacteria). Eukaryotic cells are unique because their DNA is kept in a subcompartment, the nucleus. All plants and animals, as well as yeast and many other unicellular organisms, are eucaryotes.

evolutionary psychology a field that attempts to understand the human brain and behavior on the basis of environments of the past and the effects of natural selection

foramen magnum an oval hole at the base of the skull through which the spine enters the brain; its structure is often used by paleontologists to determine whether a fossil species walked upright

forward genetics tracking down the gene or part of the genome that is responsible for a phenotype

frameshift a change in the "spelling" of a DNA or RNA molecule that shifts the borders of codons recognized by a ribosome as it translates RNA into protein

gastrulation a stage in very early embryonic development in which cell divisions and migrations lead to the formation of three layers of tissue: ectoderm, mesoderm, and endoderm

gene a region of DNA that encodes a protein

genetic drift random changes in DNA that accumulate in a population over time, eventually changing the proportion of particular forms of genes within the population

genome the entire set of DNA in an organism or species, usually referring to the DNA in nucleus (of cells that have one)

genomic technologies methods that allow researchers to monitor the behavior of a whole genome—for example, to obtain a complete readout of the genes that are active or inactive in a given type of cell, or to study the complete set of proteins that it produces

genotype an organism's complete collection of genes, including both dominant and recessive alleles

germ line related to the reproductive cells of an organism

globin a protein used to build hemoglobin and myoglobin, which transport oxygen through the body

guanine one of the nucleic acids that make up DNA and RNA molecules

haplogroup   a collection of similar haplotypes, sharing a common ancestor, grouped together by the fact that they share common DNA sequences

haplotype   short for "haploid genotype," referring to a collection of DNA sequences on the same chromosome that are inherited together

heavy chain   a large protein subunit that is used in the making of antibodies

hematopoietic   processes related to the origin and differentiation of blood cells

hemoglobin   a protein in red blood cells that binds oxygen atoms and carries them to tissues throughout the body

hermaphrodite   an organism that has the reproductive organs of both a male and female

homeobox (HOX) genes   genes which contain a code called a "homeobox," which gives them the ability to activate other genes. A subset of these genes play a crucial role in building the head-to-tail body structure of embryos.

hominid   a group of species related through evolution which includes humans, the great apes, and closely related fossil ancestors

innate immune system   the oldest form of animal immunity, including molecules and cells that recognize the presence of foreign invaders

intron   a sequence in an RNA or gene which does not encode a part of a protein

ligase   an enzyme that can join other molecules together; in genetic engineering, ligases are used to combine fragments of DNA to make new genes

light chain   a highly variable region of an antibody which participates in the recognition of foreign molecules or substances

linkage   a process by which groups of DNA sequences are inherited together

Mendelian disease   a health condition caused by inheriting dominant or recessive alleles of a single gene

mesoderm   the middle of the three main layers of tissue that arise during the process of gastrulation in early embryonic development, or tissues that arise from this layer

messenger RNA (mRNA)   a molecule that is transcribed from DNA and used as the template to create a protein

metastasis   the process by which cancer cells migrate out of a tumor, moving to other parts of the body where they create new tumors. This is usually the most deadly step in the development of cancer.

microRNA   a small RNA naturally produced by cells which does not encode a protein. Many microRNAs bind to messenger RNA molecules to block their translation into proteins.

microtubule   a fiber built of tubulin proteins that plays a key role in the transport of molecules through the cell and is used to build the mitotic spindle during cell division. Microtubules are a major part of the cytoskeleton, the scaffold of proteins that gives a cell its shape and structure.

minisatellites   small sequences of DNA that do not encode proteins but appear throughout the genome hundreds of times, with many repeated copies lying next to each other. These sequences evolve at a very rapid rate and are the basis of DNA fingerprinting.

mitochondria   organelles which provide energy for the cell. They contain their own DNA, reproduce independently of the cell nucleus, and are thought to have originated when a bacterium invaded another cell.

mitotic spindle   a structure built during cell division; it is made of microtubules, and its function is to separate chromosomes into two equal sets

molecular clock a method which attempts to establish the rate at which mutations in genes occurs, and compares different versions of molecules to determine when they (or the species they belong to) diverged

multifactorial diseases health problems caused by inheriting specific forms of multiple genes

mutagen a substance that causes a mutation in DNA

mutation a change in an organism's DNA sequence caused by a copying error, a mutagen, or some other form of damage

neocortex a brain structure in mammals that forms the outer layer of the cerebral hemispheres

nonsynonymous mutation a change in DNA that alters the amino acid spelling of a protein

olfactory receptor molecules in the membranes of neurons, responsible for the detection of odors

oncogene a gene that causes cancer if it becomes defective, often because the healthy form of the gene is involved in controlling the cell cycle

organelle a substructure within cells usually enclosed in a membrane and containing proteins and other molecules

Parkinson's disease a neurodegenerative condition caused by the death of cells in a region of the brain called the substantia nigra, leading to a loss of control of the body

penetrance the degree to which a particular form of a gene can be linked to a particular phenotype, such as a genetic disease. Complete penetrance means that the allele always produces a phenotype; partial penetrance means that people with the allele exhibit features or a disease to varying degrees.

pharmacogenetics an approach to the study and development of drugs, seeking to understand how genetic factors influence the body's responses to substances

phenotype   the complete set of measurable physical and be-havioral characteristics of an organism determined by its genes

pluripotent   the capacity of a stem cell to develop into a vari-ety of more specialized cell types

prokaryote   a species of cell that does not have a cell nucleus; this includes bacteria and archaea

protein   a molecule made of amino acids, produced by a cell based on information in its genes

receptor   a molecule in a cell or on its surface that binds to a specific partner molecule, usually changing the receptor's activity; for example, by changing the other molecules that it binds to

recessive   an allele that must be present in two copies in an organism to fully determine its phenotype

recombination   a process by which the DNA strand is cut and then the ends are rejoined at a different place, changing the order of elements in a DNA sequence as it is copied or passed down to offspring

restriction enzyme   a protein that cuts single-stranded or double-stranded DNA at specific sequences

reverse genetics   interfering with a molecule in a laboratory organism, which reveals a connection between a gene and a phenotype

reverse transcriptase   an enzyme that transcribes a single-stranded RNA molecule into a single strand of DNA, frequently used by viruses to insert their genetic information into the ge-nome of the host

Rhesus factor (Rh)   a protein used to type blood. Some people's blood cells contain Rhesus protein on their surfaces; other people's cells do not. If blood is exchanged between peo-ple with different types, the result will be a dangerous immune reaction.

RNA (ribonucleic acid) a molecule made of nucleotides that is produced by transcribing the information in a DNA sequence

single nucleotide polymorphism (SNP) a variation of a single base in the DNA sequence of two individuals

somatic mutations are changes in the genetic code that appear in some cells during an organism's lifetime (for example, through toxins or radiation damage) as opposed to germ line mutations that become part of the genetic material of every cell in a body

spandrel a concept in evolutionary theory that proposes that some features of organisms may be preserved over time without being subject to natural selection. The word comes from a technical term in architecture.

stem cell a generic cell that can differentiate into more than one specialized type over the course of development

stop codon a three-letter sequence in DNA and RNA that signals the end of the protein-encoding part of a gene

synonymous mutation a switch in a letter of a gene sequence that does not lead to any change in the chemistry of the protein it encodes

taxa groups of organisms that descend from a common ancestor

thymine one of the nucleic acids that make up DNA molecules

totipotent having the ability to develop into any type of cell in the adult body, a characteristic of embryonic stem cells

translation the process by which a ribosome reads the information in an mRNA molecule and builds a protein based on its sequence

transposon a DNA sequence that can copy itself and move from one place in the genome to another, either by being physically cut out and pasted back in somewhere else, or by first copying itself as an RNA that is then reverse transcribed into DNA

tubulin   a protein subunit used to make microtubules, part of the cell cytoskeleton

tumor suppressor gene   a gene that leads to tumors when it becomes defective; the healthy form protects cells from becoming cancerous

variable region   a part of an antibody that has a unique structure in each cell because it is created by mixing and matching random parts of genes

viral therapy   an experimental method in which genetically engineered viruses are used to deliver healthy versions of genes or RNA to patients with a disease

Further Resources

Books and Articles

Ambrose, Stanley. "Late Pleistocene Human Population Bottle-necks, Volcanic Winter, and Differentiation of Modern Humans." *Journal of Human Evolution* 34 (1998): 623–651. The paper in which Stanley Ambrose spells out his hypothesis that an eruption of a supervolcano in Indonesia nearly led to the extinction of modern humans.

Autism Research. "Autism Genetics: Strategies, Challenges, and Opportunities." Available online. URL: www3.interscience.wiley. com/cgi-bin/fulltext/117930492/HTMLSTART. Accessed March 10, 2009. A review article by Brian O'Roak and Matthew State on the search for genes related to Autism Spectrum Disorders. The article appears in the first issue of the online journal *Autism Research,* a new interdisciplinary journal (*Autism Research* 1, no. 1 (7 March 2008): 4–17.

Bodman, Walter, and Robin McKie. *The Book of Man: The Quest to Discover Our Genetic Heritage.* London: Little, Brown and Company, 1994. A well-written account of the major people and themes of human genetics from the late 19th century to the beginning of the Human Genome Project.

Branden, Carl, and John Tooze. *Introduction to Protein Structure.* 2d ed. New York: Garland Publishing, 1999. A detailed overview of the chemistry and physics of proteins, for university students with some background in both fields.

Brown, Andrew. *In the Beginning Was the Worm.* London: Pocket Books, 2004. The story of an unlikely model organism in biology: the worm *C. elegans,* and the scientists who have used it to understand some of the most fascinating issues in modern biology.

Browne, Janet. *Charles Darwin: The Power of Place.* New York: Knopf, 2002. The second volume of the definitive biography of Charles Darwin.

————. *Charles Darwin: Voyaging.* Princeton, N.J.: Princeton University Press, 1995. The first volume of the definitive biography of Charles Darwin.

Brunet, Michel, et. al. "A New Hominid from the Upper Miocene of Chad, Central Africa." *Nature* 418 (2002): 145–51. The original paper reporting the find of *Sahelanthropus tchadensis,* a 7-million-year-old hominid that may be closely related to the common ancestor of humans and chimpanzees.

Cann, Rebecca, Mark Stoneking, and Allan C. Wilson. "Mitochondrial DNA and Human Evolution." *Nature* 325 (1987): 31–36. The original paper using mitochondrial DNA sequences to trace modern human ancestry to "Mitochondrial Eve."

Caporale, Lynn Helena. *Darwin in the Genome: Molecular Strategies in Biological Evolution.* New York: McGraw-Hill, 2003. A new look at variation and natural selection based on discoveries from the genomes of humans and other species, written by a noted biochemist.

Carlson, Elof Axel. *Mendel's Legacy: The Origin of Classical Genetics.* Cold Spring Harbor, N.Y.: Cold Spring Harbor Laboratory Press, 2004. An excellent, easy-to-read history of genetics from Mendel's work to the 1950s. Carlson explains the relationship between cell biology and genetics especially well.

————. *The Unfit: A History of a Bad Idea.* Cold Spring Harbor, N.Y.: Cold Spring Harbor Laboratory Press, 2001. An in-depth account of eugenics movements across the world.

Cavalli-Sforza, L. Luca. "The Human Genome Diversity Project: Past, Present, and Future." *Nature Reviews Genetics* 6 (2005): 333. An overview of the progress and difficulties encountered by the HGDP from its conception to its development as a resource now being used by researchers all over the world.

Cavalli-Sforza, L. Luca, Paolo Menozzi, and Alberto Piazza. *The History and Geography of Human Genes.* Princeton, N.J.:

Princeton University Press, 1994. For over three decades Cavalli-Sforza has been interested in using genes (as well as other fields such as linguistics) to study human diversity and solve interesting historical questions, such as where modern humans evolved and how they spread across the globe. This book is a compilation of what he and many other researchers have found.

Chambers, Donald A. *DNA: The Double Helix: Perspective and Prospective at Forty Years.* New York: New York Academy of Sciences, 1995. A collection of historical papers from major figures involved in the discovery of DNA, with reminiscences from some of the authors.

Chen, Feng-Chi, and Wen-Hsiung Li. "Genomic Divergences between Humans and Other Hominoids and the Effective Population Size of the Common Ancestor of Humans and Chimpanzees." *American Journal of Human Genetics* 68 (2001): 444–56. A review that uses a "molecular clock" methodology to estimate when humans and chimpanzees diverged from a common ancestor.

The Chimpanzee Sequencing and Analysis Consortium. "Initial sequence of the chimpanzee genome and comparison with the human genome." *Nature* 437 (2005): 69–87. This article presents an in-depth contrast of the complete DNA sequences of humans and chimpanzees.

Crick, Francis. *What Mad Pursuit: A Personal View of Scientific Discovery.* New York: Basic Books, 1988. Crick's account of dead ends, setbacks, wild ideas, and finally glory on the road to the discovery of the structure of DNA, with speculations on the future of neurobiology and other fields.

Darwin, Charles. *The Descent of Man.* Amherst, N.Y.: Prometheus, 1998. In this book, originally published 12 years after *On the Origin of Species,* Darwin outlines his ideas on the place of human beings in evolutionary theory.

———. *On the Origin of Species.* Edison, N.J.: Castle Books, 2004. Darwin's first, enormous work on evolution, which examines a huge number of facts while building a case for heredity, variation, and natural selection as the forces that produce new species from existing ones.

———. The *Voyage of the Beagle.* London: Penguin Books, 1989. A scientific adventure story; Darwin's account of his five years as a young naturalist aboard the *Beagle.* He had not yet discovered the principles of evolution but was aware of the need for a scientific theory of life. Readers watch over his shoulder as he tries to make sense of questions that puzzled scientists everywhere in the mid-19th century.

Diamond, Jared. *Guns, Germs, and Steel.* New York: W. W. Norton, 2005. A new look at how Western civilization came to dominate the globe, integrating information from archeology, anthropology, genetics, evolutionary biology, and many other sources. Many consider Diamond the first to accurately apply evolutionary principles to the question of why some societies become dominant over others.

Dorus, Steve, Eric J. Vallender, Patrick D. Evans, Jeffrey R. Anderson, Sandra L. Gilbert, Michael Mahowald, Gerald J. Wyckoff, Christine M. Malcom, and Bruce T. Lahn. "Accelerated Evolution of Nervous System Genes in the Origin of *Homo sapiens." Cell* 119 (7) (December 29, 2004). 1027–40. A systematic study of how quickly genes crucial to brain evolution have evolved in primates and mammals. The study compares genes in humans, macaque monkeys, mice, and rats, revealing that brain-related genes have been subject to pressure from natural selection, especially in the branch of primates leading to humans.

Dubrova, Yuri, Valeri Nesterov, Nicolay Krouchinsky, Vladislav Ostapenko, Gilles Vergnaud, Fabienne Giraudeau, Jérome Buard, and Alec Jeffreys. "Further Evidence for Elevated Human Minisatellite Mutation Rate in Belarus Eight Years after the Chernobyl Accident." *Mutation Research* 381 (2007): 267–78. A groundbreaking study examining the long-term impact of the Chernobyl accident on the genomes of people living in the region.

Elliott, William H., and Daphne C. Elliott. *Biochemistry and Molecular Biology.* New York: Oxford University Press, 1997. An excellent college-level overview of the biochemistry of the cell.

Fruton, Joseph. *Proteins, Enzymes, Genes: The Interplay of Chemistry and Biology.* New Haven, Conn.: Yale University Press, 1999. A very detailed historical account of the lives and work of the chemists, physicists, and biologists who worked out the major functions of the molecules of life.

Gilbert, Scott. *Developmental Biology.* Sunderland, Mass.: Sinauer Associates, 1997. An excellent college-level text on all aspects of developmental biology.

Goldsmith, Timothy H., and William F. Zimmermann. *Biology, Evolution, and Human Nature.* New York: Wiley, 2001. Life from the level of genes to human biology and behavior.

Gregory, T. Ryan, ed. *The Evolution of the Genome.* Boston: Elsevier Academic Press, 2005. An advanced-level book presenting the major themes of evolution in the age of genomes, written by leading researchers for graduate students and scientists.

Harper, Peter S. *Practical Genetic Counselling.* 6th ed. London: Hodder Arnold, 2004. An introduction to the basics of genetic counseling, with a review of genetic disorders based on body systems such as the nervous system, the eye, and cardiovascular and respiratory diseases, including a section on privacy issues and other themes under the heading "Genetics and Society."

Henig, Robin Marantz. *A Monk and Two Peas.* London: Weidenfeld & Nicolson, 2000. A popular, easy-to-read account of Gregor Mendel's work and its impact on later science.

Herrnstein, Richard J., and Charles Murray. *The Bell Curve: Intelligence and Class Structure in American Life.* New York: Free Press, 1996. A controversial examination of the connection between genes and IQ, in which the authors attempt to connect heredity to overall intelligence and draw social conclusions about themes such as how educational systems should be organized.

Horiike, Tokumasa, Kazuo Hamada, Shigehiko Kanaya, and Takao Shinozawa. "Origin of Eukaryotic Cell Nuclei by

Symbiosis of Archaea in Bacteria Is Revealed by Homology-hit Analysis." *Nature Cell Biology* (2001): 210–14. A study comparing a wide range of genes from eukaryotes, bacteria, and archaea, which revealed that the nucleus of eukaryotic cells may have originally been an archaeal cell that took up residence in a bacterium.

Judson, Horace Freeland. *The Eighth Day of Creation: Makers of the Revolution in Biology.* New York: Simon and Schuster, 1979. A comprehensive history of the science and people behind the creation of molecular biology, from the early 20th century to the 1970s, based on hundreds of hours of interviews Judson conducted with the researchers who created this field.

Kaas, Jon. "The Evolution of the Complex Sensory and Motor Systems of the Human Brain." *Brain Research Bulletin* 75 (2008): 384–90. This review collects data taken from casts made of fossil skulls and living animals to induce how functional modules developed as humans evolved from primates.

Keller, Evelyn Fox. *A Feeling for the Organism: The Life and Work of Barbara McClintock.* San Francisco: W.H. Freeman, 1983. An account of the life and work of the discoverer of jumping genes, written before she received the Nobel Prize for her work. The book reveals the problems encountered by a brilliant, radical thinker—as well as a woman—working in science during the middle of the 20th century.

Koch, Christof. *The Quest for Consciousness: A Neurobiological Approach.* Englewood, Colo.: Roberts & Company, 2004. Koch is trying to establish the biological basis of consciousness and related mental abilities in humans. This fascinating book explores discoveries from the activity of genes to the behavior of modules of the brain and presents scientists' best current knowledge of the relationship between the physical brain and the metaphysical mind.

Kohler, Robert E. *Lords of the Fly: Drosophila Genetics and the Experimental Life.* Chicago: University of Chicago Press, 1994. The story of Thomas Hunt Morgan and his "disciples," whose discoveries regarding fruit fly genes dominated genetics in the first half of the 20th century.

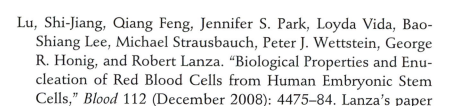

Lu, Shi-Jiang, Qiang Feng, Jennifer S. Park, Loyda Vida, Bao-Shiang Lee, Michael Strausbauch, Peter J. Wettstein, George R. Honig, and Robert Lanza. "Biological Properties and Enucleation of Red Blood Cells from Human Embryonic Stem Cells," *Blood* 112 (December 2008): 4475–84. Lanza's paper describing the successful creation of differentiated blood from embryonic stem cells in cell cultures in the laboratory.

Lutz, Peter L. *The Rise of Experimental Biology: An Illustrated History.* Totowa, N.J.: Human Press, 2002. A very readable, wonderfully illustrated book tracing the history of biology from ancient times to the modern era.

Maddox, Brenda. *Rosalind Franklin: The Dark Lady of DNA.* London: HarperCollins Publishers, 2002. An account of the life and work of Rosalind Franklin, who played a key role in the discovery of DNA's structure but who had trouble fitting in to the scientific culture of London in the 1950s.

Magner, Lois N. *A History of the Life Sciences.* New York: Marcel Dekker, 1979. An excellent, wide-ranging book on the development of ideas about life from ancient times to the dawn of genetic engineering.

McElheny, Victor K. *Watson and DNA: Making a Scientific Revolution.* Cambridge, Mass.: Perseus, 2003. A retrospective on the work and life of the extraordinary scientific personality James Watson.

Ptashne, Mark, and Gann, Alexander. *Genes and Signals.* Cold Spring Harbor, N.Y.: Cold Spring Harbor Laboratory Press, 2002. A readable and nicely illustrated book presenting a modern view of how genes in bacteria are regulated and what these findings mean for the study of other organisms.

Purves, William K., David Sadava, Gordon H. Orians, and Craig Heller. *Life: The Science of Biology.* Kenndallville, Ind.: Sinauer Associates and W. H. Freeman, 2003. A comprehensive overview of themes from the life sciences.

Reeve, Eric, ed. *Encyclopedia of Genetics.* London: Fitzroy Dearborn Publishers, 2001. A very interesting collection of essays on the major themes in genetics by researchers who are world leaders in their fields. A significant section is devoted to forensic studies of famous cases such as that of Anastasia, the Jefferson family, and Jesse James.

Sacks, Oliver. *The Island of the Colour-Blind and Cycad Island.* London: Picador, 1996. Neurobiologist Sacks's personal account of his travels to the Pacific islands of Cycad and Pingelap. There he encountered people with an unusual genetic condition that allows them only to see shades of gray; the book contains his reflections on the impact of this condition on island culture.

Scott, Christopher. *Stem Cell Now: From the Experiment that Shook the World to the New Politics of Life.* New York: Pi Press, 2006. An excellent, very readable introduction to stem cells and the role that they are likely to play in the medicine of the future, taking into account political, social, and ethical dimensions of their use.

Skaletsky, Helen, et al. "The male-specific region of the human Y chromosome is a mosaic of discrete sequence classes." *Nature* 423 (2003): 825–38. A fascinating article tracing the evolutionary origins of the genes on the human Y chromosome.

Stent, Gunther. *Molecular Genetics: An Introductory Narrative.* San Francisco: W.H. Freeman, 1971. A classic book for college-level students about the development of genetics and molecular biology by a researcher and teacher who witnessed it firsthand.

Strachan, Tom, and Andrew P. Read. *Human Molecular Genetics 3.* New York: Garland Publishing, 2004. An excellent college-level textbook giving a comprehensive overview of methods and findings in human genetics in the molecular age.

Sykes, Bryan. *The Seven Daughters of Eve: The Science that Reveals Our Genetic Ancestry.* New York: W.W. Norton & Company, Inc., 2001. Sykes is one of the pioneers in the use of mitochondrial DNA to trace the origins of modern humans and their migrations across the globe. This book is a very entertaining and personal account of his discoveries and work, including his analysis of the path taken by seafarers as they settled the islands of the Pacific, and studies of the origins of the populations of Europe.

Tanford, Charles, and Jacqueline Reynolds. *Nature's Robots: A History of Proteins.* New York: Oxford University Press Inc., 2001. A history of biochemical and physical studies of proteins and their functions and the major researchers in the field.

Treffert, Darold. *Extraordinary People: Understanding Savant Syndrome.* New York: Backinprint.com, 2006. A scientific, personal, and historical account of savants from a psychiatrist who has worked with and studied them for several decades.

Trinkaus, Erik. "European Early Modern Humans and the Fate of the Neandertals." *Proceedings of the National Academy of Sciences* 104, no. 18 (2007): 7367–72. A study by one of the world's foremost experts on Neanderthals, who maintains that they may have interbred with modern humans that settled Europe.

Tudge, Colin. *In Mendel's Footnotes.* London: Vintage, 2002. An excellent review of ideas and discoveries in genetics from Mendel's day to the 21st century.

———. *The Variety of Life: A Survey and a Celebration of All the Creatures that Have Ever Lived.* New York: Oxford University Press, 2000. A beautifully illustrated "tree of life" classifying and describing the spectrum of life on Earth.

Vogel, Friedrich, and Arno Motulsky. *Human Genetics.* 3d ed. New York: Springer-Verlag, 1997. A college-level, in-depth review of human genetics in the molecular age.

Wang, Eric, Greg Kodama, Pierre Baldi, and Robert K. Moyzis. "Global Landscape of Recent Inferred Darwinian Selection for *Homo sapiens.*" *Proceedings of the National Academy of Sciences* 103, no. 1 (2006): 135–40. A study comparing DNA sequences from humans, apes, and rodents. The work reveals human genes that have been subject to natural selection, particularly genes related to brain development and function.

Watson, James D. *The Double Helix.* New York: Atheneum, 1968. Watson's personal account of the discovery of the structure of DNA.

Watson, James D., and Francis Crick. "A Structure for Deoxyribose Nucleic Acid." *Nature* 171 (1953): 737–38. The original article in which Watson and Crick described the structure of DNA and its implications for genetics and evolution.

Way, Gregory. "Dating Branches on the Tree of Life Using DNA." *Genome Biology* 3, no. 1 (2001): reviews 0001.1. This article from an online journal of biocomputing reviews how "molecular clock" methods are being used to date key events in evolution.

Wolpoff, Milford, et al. "An Ape or *the* Ape: Is the Toumaï Cranium TM 266 a Hominid?" *PaleoAnthropology* 2006: 36–50. This article reviews the evidence that *Sahelanthropus tchadensis,* a 7-million-year-old fossil discovered by Michel Brunet, might be closely related to the common ancestor of humans and chimpanzees. The article gives a fascinating look at how paleoanthropologists interpret teeth and bones. Also available online at www.paleoanthro.org/journal/content/PA20060036.pdf

Web Sites

There are tens of thousands of Web sites devoted to the topics of molecular biology, genetics, evolution, and the other themes of this book. The small selection below provides original articles, teaching materials, multimedia resources, and links to hundreds of other excellent sites.

The American Society of Naturalists. "Evolution, Science, and Society: Evolutionary Biology and the National Research Agenda." Available online. URL: http://www.rci.rutgers.edu/~ecolevol/fulldoc.pdf. Accessed March 10, 2009. A document from the American Society of Naturalists and several other organizations, summarizing evolutionary theory and showing how it has contributed to other fields including health, agriculture, and the environmental sciences.

The Bradshaw Foundation. "Journey of Mankind—The Peopling of the World." Available online. URL: www.bradshaw

foundation.com/journey. Accessed March 10, 2009. An online lecture and film giving an excellent visual demonstration of how and when modern humans likely spread from Africa to populate the globe.

The California Institute of Technology. "The Caltech Institute Archives." Available online. URL: archives.caltech.edu/index.cfm. Accessed March 10, 2009. This site provides materials tracing the history of one of America's most important scientific institutes since 1891. One highlight is a huge collection of oral histories with firsthand accounts of some of the leading figures who have been at Caltech, including George Beadle and Max Delbrück.

Center for Evolutionary Psychology. "Evolutionary Psychology Primer by Leda Cosmides and John Tooby." Available online. URL: www.psych.ucsb.edu/research/cep/primer.html. Accessed March 10, 2009. A basic, easy-to-understand introduction to the concepts and principles of evolutionary psychology by two of the foremost researchers in the field.

The Center for Genetics and Society. "CGS: Detailed Survey Results." Available online. URL: www.geneticsandsociety.org/article.php?id=404. Accessed March 10, 2009. This article presents the results of numerous surveys conducted in the United States and elsewhere on topics related to genetics, human cloning, and stem cell research, providing a fascinating view of people's knowledge of basic genetic topics as well as how opinions have changed over the past few years.

Centers for Disease Control and Prevention. "HPV Vaccine Information for Young Women." Available online. URL: www.cdc.gov/std/hpv/STDFact-HPV-vaccine.htm. Accessed March 10, 2009. An overview of issues regarding the human papilloma virus and vaccines providing valuable information, particularly for young women, about the risk of infection and useful information about the vaccine.

The Dolan DNA Learning Center, Cold Spring Harbor Laboratory. "DNA Interactive." Available online. URL: http://www.dnai.org. Accessed March 10, 2009. A growing collection of multimedia and archival materials including several hours of filmed interviews with leading figures in molecular biology, a timeline of discoveries, an archive on the American eugenics movement, and a wealth of teaching materials on the topics of this book.

————. "Genes to Cognition Online." Available online. URL: www.g2conline.org. Accessed March 10, 2009. This Web site for students, teachers, and the general public (as well as scientists) offers a huge amount of material on the relationship between genes and thinking and a wide range of related topics. A unique feature of the site is a new, dynamic style of navigation based on "concept mapping," a learner-directed technique for structuring and visualizing information. The DNA Learning Center is currently testing the site in classrooms to explore new ways of teaching and learning about science.

The European Bioinformatics Institute. "2can." Available online. URL: www.ebi.ac.uk/2can/home.html. Accessed March 10, 2009. An educational site from the EBI—one of the world's major Internet providers of information about genomes, proteins, molecular structures, and other types of biological data. Many of the tutorials and basic introductions to the themes are accessible to pupils or people with a bit of basic knowledge in biology.

The Exploratorium. "Microscope Imaging Station." Available online. URL: www.exploratorium.edu/imaging_station/index. php. Accessed March 10, 2009. San Francisco's Exploratorium is an interactive science museum; its Web site has a range of wonderful activities based on biological themes such as development, blood, stem cells, and the brain. There are also videos, desktop wallpapers that can be downloaded for free, and feature articles on current themes from science.

Human Genome Project, U.S. Department of Energy. "Genetics Legislation." Available online. URL: www.ornl.gov/sci/techresources/Human_Genome/elsi/legislat.shtml. Accessed March 10, 2009. This page presents an overview of legislation regarding human genome information and the protection of personal genetic information. The Web site is a good starting point for teachers and students who want to get an overview of scientific and ethical issues related to human genetics, including information about laws pertaining to genetic testing, patient rights, medical discoveries, and other topics.

The Institute of Human Origins. "Becoming Human: Paleoanthropology, Evolution and Human Origins." Available online. URL: www.becominghuman.org. Accessed March 10, 2009. An attractive site with a focus on paleoanthropology and human origins, with a video documentary that can be watched online or downloaded, classroom resources, and articles on "How Science Is Done." "The Chromosome Connection," an activity in the Learning Center section of the site, introduces pupils to differences between humans and apes from a molecular perspective.

Maddison, David R., and Katja-Sabine Schulz, eds. "The Tree of Life Web Project." Available online. URL: tolweb.org. Accessed March 10, 2009. A site that has collected a huge number of articles and links from noted biologists on the question of assembling a "family tree" of life on Earth.

The National Center for Biotechnology Information. "Bookshelf." Available online. URL: www.ncbi.nlm.nih.gov/sites/entrz?db=books. Accessed March 10, 2009. A collection of excellent online books ranging from biochemistry and molecular biology to health topics. Most of the works are quite technical, but many include very accessible introductions to the topics. Some highlights are *Molecular Biology of the Cell, Molecular Cell Biology,* and the *Wormbook.* There are also annual reports on health in the United States from the Centers for Disease Control and Prevention.

National Geographic. "Outpost: Human Origins@nationalgeographic.com" Available online. URL: www.nationalgeographic.com/features/outpost. Accessed March 10, 2009. A virtual expedition, accompanying human fossil hunter Lee Berger on a search for ancient human remains in Botswana and South Africa.

The National Health Museum. "Access Excellence: Genetics Links." Available online. URL: www.accessexcellence.org/RC/genetics.php. Accessed March 10, 2009. Links and resources from the "Access Excellence" project of the National Health Museum, in Atlanta.

National Public Radio. "Wild Cows Cloned." Available online. URL: www.npr.org/templates/story/story. php?storyId=1225049. Accessed March 10, 2009. An audio interview with researcher Robert Lanza, pioneering stem cell researcher (see chapter 1, "Making Blood in the Lab"). In this interview Lanza discusses his cloning of an endangered wild cow called the banteng, using cells taken from an animal that had died 23 years earlier. Lanza also appears in an interview from 2006 on more general themes in stem cell research at the following URL: www.npr.org/templates/story/ story.php?storyId=5204335, accessed March 10, 2009. The NPR Web site offers a wide range of interviews and broadcasts on biological and medical themes that can be heard online.

The Nobel Foundation. "Video Interviews with Nobel Laureates in Physiology or Medicine." Available online. URL: nobel prize.org/nobel_prizes/medicine/video_interviews.html. Accessed March 10, 2009. Video interviews with laureates from the past four decades, many of whom have been molecular biologists or researchers from related fields. Follow links to interviews with winners of other prizes, Nobel lectures, and other resources.

The Public Broadcasting Service. "American Experience: Jesse James." Available online. URL: www.pbs.org/wgbh/amex/ james/index.html. Accessed March 10, 2009. This is the home page of a PBS documentary centered on the life of Jesse James, including the transcript of the broadcast, a wide range of images, and other supplementary materials.

The Research Collaboratory for Structural Bioinformatics. "RCSB Protein Data Bank." Available online. URL: www. rcsb.org. Accessed March 10, 2009. This site provides "a variety of tools and resources for studying the structures of biological macromolecules and their relationships to sequence, function, and disease." There is a multimedia tutorial on how to use the tools and databases. One special feature is the "Molecule of the Month," with beautiful illustrations by David Goodsell.

The School of Crystallography of Birkbeck College, the University of London. "The Principles of Protein Structure." Available online. URL: www.cryst.bbk.ac.uk/PPS. Accessed March 10, 2009. An online university program through which students can enroll to study structural biology on the Internet. The site hosts a wide range of basic introductory materials explaining the fundamentals of the field.

Science Friday. "Science Friday Archives: Cancer Update with Robert Weinberg." Available online. URL: www.science friday.com/program/archives/200710123. Accessed March 10, 2009. Science Friday is heard live every Friday on National Public Radio stations. This link points to the podcast of a program originally broadcast on October 12, 2007, with Robert Weinberg, renowned cancer researcher at MIT. Weinberg's group had just discovered that small molecules called micro RNAs regulate the production of proteins in tumor cells, a finding with significant implications for diagnosis and therapy.

Scientific American. "The First Human Cloned Embryo." Available online. URL: www.sciam.com/article.cfm?id=the-first-human-cloned-em. Accessed March 10, 2009. An article by Robert Lanza and his colleagues describing the methods they used to make the first clones of human embryonic stem cells. The article explains the methods used and the researchers' hopes for how this type of work may change medicine in the future.

TalkOrigins. "The TalkOrigins Archive." Available online. URL: www.talkdesign.org. Accessed March 10, 2009. A Web site devoted to "assessing the claims of the Intelligent Design movement from the perspective of mainstream science; addressing the wider political, cultural, philosophical, moral, religious, and educational issues that have inspired the ID movement; and providing an archive of materials that critically examine the scientific claims of the ID movement." A subsection of the site deals specifically with the origins of humans: www.talkorigins.org/faqs/homs.

The Tech Museum of Innovation, San Jose, Calif. "Understanding Genetics: Human Health and the Genome." Available online. URL: www.thetech.org/genetics. Accessed March 10, 2009. An excellent collection of news and feature stories on scientific discoveries and ethical issues surrounding genetics.

University of California at Santa Cruz. "UCSC Genome Bioinformatics." Available online. URL: genome.ucsc.edu/bestlinks. html. Accessed March 10, 2009. A portal to high-quality resources for the study of molecules and genomes, from UCSC and other sources. The map of the BRCA2-gene presented in chapter 2 was obtained using this site.

University of Cambridge. "The Complete Works of Charles Darwin Online." Available online. URL: darwin-online.org. uk. Accessed March 10, 2009. An online version of Darwin's complete publications, 20,000 private papers, and hundreds of supplementary works.

University of Utah Genetic Science Learning Center. "Learn Genetics." Available online. URL: learn.genetics.utah.edu. Accessed March 10, 2009. An excellent Web site introducing the basics of genetic science, including a "Biotechniques Virtual Laboratory," special features on the genetics and neurobiology of addiction, stem cells, and molecular genealogy, and podcasts on the genetics of perception and aging.

University of Washington Television. "UWTV Program: Genomic Views of Human History." Available online. URL: www. uwtv.org/programs/displayevent.aspx?rID=2493. Accessed March 10, 2009. A lecture from Mary-Claire King (chapter 4) that can be watched online. The theme is how new tools of genomic analysis are being used to investigate the genes of modern humans, shedding light on historical puzzles such as ancient migrations and the settlement of the globe.

The Vega Science Trust. "Scientists at Vega." Available online. URL: www.vega.org.uk/video/internal/15. Accessed March 10, 2009. Filmed interviews with some of the great figures in 20th-century and current science, including Max Perutz,

Kurt Wüthrich, Aaron Klug, Fred Sanger, John Sulston, Bert Sakmann, Christiane Nüsslein-Volhard, and others.

The Wisconsin Medical Society home page. Available online. URL: www.wisconsinmedicalsociety.org/savant_syndrome. Accessed March 10, 2009. This site is devoted to the work of Darold Treffert, a psychiatrist who is likely the world's foremost expert on people with the syndrome. It includes an overview of the field, references to the literature, and written and video portraits of over 20 savants.

Index

7/18
T=1 9/10
L=2

7/1
T 1
9/10
L 2

1/16
T-1 9/10
L-2

6/13
T 1
9/10
L 2

GAYLORD